Neutron Diffusion
Concepts and Uncertainty Analysis for Engineers and Scientists

Neutron Diffusion
Concepts and Uncertainty Analysis for Engineers and Scientists

S. Chakraverty and Sukanta Nayak

CRC Press
Taylor & Francis Group
Boca Raton London New York

CRC Press is an imprint of the
Taylor & Francis Group, an **informa** business

CRC Press
Taylor & Francis Group
6000 Broken Sound Parkway NW, Suite 300
Boca Raton, FL 33487-2742

First issued in paperback 2019

© 2017 by Taylor & Francis Group, LLC
CRC Press is an imprint of Taylor & Francis Group, an Informa business

No claim to original U.S. Government works

ISBN-13: 978-1-4987-7876-3 (hbk)
ISBN-13: 978-0-367-88980-7 (pbk)

Library of Congress Cataloging-in-Publication Data

Names: Chakraverty, Snehashish. | Nayak, Sukanta.
Title: Neutron diffusion : concepts and uncertainty analysis for engineers and scientists / S. Chakraverty and Sukanta Nayak.
Description: Boca Raton : CRC Press, 2017. | Includes bibliographical references.
Identifiers: LCCN 2016041880 | ISBN 9781498778763 (hardback) | ISBN 9781498778770 (ebook)
Subjects: LCSH: Neutron transport theory. | Differential equations.
Classification: LCC QC793.5.N4622 C43 2017 | DDC 539.72/13–dc23
LC record available at https://lccn.loc.gov/2016041880

Visit the Taylor & Francis Web site at
http://www.taylorandfrancis.com

and the CRC Press Web site at
http://www.crcpress.com

Contents

Preface

The design of the reactor core is based on the description of the production, transport and absorption of neutrons. As neutrons move within a medium, viz. gas, liquid or solid, they collide with the nuclei of the atoms in the medium. During such collisions, neutrons may be absorbed by the nuclei, which are either elastic or inelastic. Absorption of neutrons may result in a loss or an increase in the number of neutrons by fission. Fission neutrons usually have different energies and move in different directions than incident neutrons. As a result, there may be scattering of neutrons, which changes the position, energy and direction of the motion of the neutrons. Furthermore, the scattering of neutron collision inside a reactor depends upon the geometry of the reactor, diffusion coefficient and absorption coefficient. It may also be noted that the parameters responsible for the diffusion (i.e. the scattering of neutron) may not always be crisp, rather they may be uncertain.

In general, these uncertainties occur due to the vague, imprecise and incomplete information about the variables and parameters, as a result of errors in measurement, observation, experiment or applying different operating conditions or due to maintenance-induced errors, which are uncertain in nature. So, to overcome these uncertainties, one may use the fuzzy/interval or stochastic environmental parameters and variables in place of crisp (fixed) parameters. With these uncertainties, the governing differential equations turn fuzzy/ interval or stochastic. Practically, it is sometimes difficult to obtain the solution of fuzzy equations due to the complexity in the fuzzy arithmetic; for example, addition and multiplication are not the inverse operations of subtraction and division, respectively. On the other hand, we may model the problem as stochastic only when sufficient data are available. Moreover, the combination (hybridization) of fuzzy and stochastic is also a new vista. These were found to be important and challenging areas of study in recent years. As such, one may need to understand the nuclear diffusion principles/theories corresponding to reliable and efficient techniques for the solution of such uncertain problems.

Accordingly, the objective of this book is to provide first the basic concepts of reactor physics as well as neutron diffusion theory. The main aim of the book, however, is about handling uncertainty in neutron diffusion problems. Hence, uncertainties (i.e. fuzzy, interval, stochastic) and their applications in nuclear diffusion problems have been included here in a systematic manner, along with the recent developments. This book may be an essential reference for students, scholars, practitioners, researchers and academicians in the assorted fields of engineering and science, particularly nuclear engineering.

Chapter 1 describes the preliminaries of basic reactor principles. Here, a few important and related terminologies for the nuclear reactor are briefly explained. In Chapter 2, neutron diffusion theory and the formulation of the neutron diffusion equation have been presented, which give first-hand scientific insights to study various nuclear design problems. The transportation of the scattered neutrons is modelled and formulated mathematically. In Chapter 3, the fundamentals of uncertainties are discussed. In this chapter, uncertainties are addressed with respect to three categories, viz. interval, fuzzy and probabilistic. The operations of these uncertainties are demonstrated through various examples. Furthermore, Chapter 4 elaborates the uncertain modelling (considering interval/fuzzy parameters as uncertain) of neutron diffusion. The factors involved in the reactor system, which are responsible for uncertainness, are modelled in terms of interval/fuzzy.

One-group models with respect to crisp parameters are explained in Chapter 5. This chapter includes both the analytical and numerical approaches to investigate one-group models, whereas in Chapter 6, uncertain (considering interval/fuzzy parameters as uncertain) one-group models are discussed, and example problems are investigated. In this chapter, the uncertain neutron diffusion equation for a bare square homogeneous reactor is discussed, which has been modelled by the fuzzy finite element method. The multigroup neutron diffusion equation has been generalized in Chapter 7, and again the finite element method has been used to solve the same. In Chapter 8, the uncertain multigroup neutron diffusion model is investigated. Accordingly, a benchmark problem is solved under an uncertain environment, and the sensitivity of the uncertain parameters is also analyzed.

Chapter 9 includes the theory of point kinetic diffusion. Here, different terminologies related to point kinetic diffusion of the one-group bare reactor are discussed, and the point kinetic diffusion equation for crisp parameters is formulated. Next, the stochastic point kinetic diffusion equation is modelled in Chapter 10. The basic concept of the birth–death stochastic process and stochastic point kinetic diffusion are discussed here. Finally, in Chapter 11, hybridized uncertainty (i.e. both probabilistic and fuzzy) is considered, and uncertain point kinetic diffusion is modelled. Furthermore, the hybridized uncertainty for this model is demonstrated through an example problem.

This book aims to provide a systematic understanding of nuclear diffusion theory along with uncertainty, viz. fuzzy/interval/stochastic and hybrid methods. The book will certainly prove to be a benchmark for graduate and postgraduate students, teachers, engineers and researchers in this field. It provides comprehensive results and an up-to-date and self-contained review of the topic along with application-oriented treatment of the use of newly developed methods of different fuzzy, stochastic and hybrid computations in various domains of nuclear engineering and sciences. It is worth mentioning that the presented methods may very well be used in/extended to various other engineering disciplines, viz. electronics, marine, chemical and mining, and sciences such as physics, chemistry and biotechnology, where one wants to model physical problems with respect to non-probabilistic (interval/fuzzy) and hybrid uncertainties for understanding the real scenario.

S. Chakraverty
Sukanta Nayak

1

Basic Reactor Principles

The principle of the nuclear reactor is based on the transport of neutrons and their interaction with matter within a reactor. Various terminologies involved in the design of nuclear reactors, based on the fission process, have been included in this chapter.

1.1 Atomic Structure

An atom is a small part of an element which consists of a positively charged nucleus surrounded by negatively charged *electrons*. Atomic nuclei are made up of two fundamental particles, viz. protons and neutrons. The *proton* carries a unit positive charge (+e) and a mass about 1836 times the electronic mass (m_e). In fact, the proton is identical with the nucleus of a hydrogen atom. On the other hand, the *neutron* is electrically neutral and slightly heavier than the proton. The neutrons and protons are together called the *nucleons*.

For each element, protons represent the atomic number of the atomic nucleus and are denoted by the symbol Z. It is also the same as the ordinal number of the element in the familiar periodic table of the elements. The total number of neutrons (N) and protons (Z) inside the nucleus is called the *mass number* (A), that is $A = N + Z$. In general, a nucleus of an atom X of atomic number Z and mass number A is symbolically written as $_Z^A X$. For example, the helium atom is denoted by $_2^4 He$, with the atomic number 2 and mass number 4. These are also known as α-particles. The protons and neutrons are symbolically denoted as p and n, respectively. The subscript to the left (i.e. atomic number) of the atomic symbol is often omitted and it looks like $^A X$.

Nuclei with the same atomic number (Z) but a different mass number (A) are called *isotopes*. A particular element with a given Z may have isotopes of different mass numbers. Their nuclei possess the same number of protons, whereas a different number of neutrons. For example, carbon has the isotopes ^{12}C, ^{13}C and ^{14}C with mass numbers 12, 13 and 14, respectively. On the other hand, nuclei with the same mass number, but different atomic numbers, are known as *isobars*, while those with the same number of neutrons are known as *isotones*. Chlorine and potassium are isotones that have 20 neutrons in their nuclei.

The mass of each atom is expressed in terms of the atomic mass unit (u), which is exactly 1/12 of the mass of the ^{12}C atom. Uranium is the most important element for the release of nuclear energy. In nature, it exists in three different forms with mass numbers 134, 235 and 238, respectively. Apart from this, another important element is thorium, with the atomic number 90 and mass number 232 (Hetrick 1971).

1.2 Binding Energy

In nuclear physics, the nuclear masses are determined by using the mass spectroscopes (Ghoshal 2010). The actual mass is always less than the sum of the masses of the constituent nucleons. The difference is called the *mass defect*. The mass defect is related to the energy binding the particles in the nucleus. In other words, if we want to break up a nucleus of Z protons and N neutrons completely so that they are all separated from each other, a certain minimum amount of energy is to be supplied to the nucleus. This energy is known as the *binding energy*.

Let us consider the mass of protons, nucleus and electrons as m_p, m_n and m_e, respectively. The mass defect is defined as (Stacey 2007)

$$\text{Mass defect} = \left[Z\left(m_p + m_e\right) + \left(A - Z\right)m_n \right] - M,$$

where
$Z(m_p + m_e) + (A - Z)m_n$ is the masses of the constituents of the atom
M is the observed mass of the atom

Using the Einstein equation, the binding energy can be calculated as follows:

$$E = \Delta mc^2 \tag{1.1}$$

where
Δm is the mass defect
c is the velocity of light ($\approx 3 \times 10^8$ m/s)

In other words, binding energy = $\Delta m \times 931.5$ MeV, where 1 u = 931.5 MeV.

1.2.1 Calculation of 1 u

It is well known that 1 mole of ^{12}C has the mass of 12 g or 12×10^{-3} kg. Since 1 mole contains 6.02205×10^{23} atoms (Avogadro number), the mass of each ^{12}C atom is

$$\frac{12 \times 10^{-3}}{6.02205 \times 10^{23}} \text{ kg} = 12 \times 1.66056 \times 10^{-27} \text{ kg}.$$

Hence, the unit of atomic mass is

$$1 \text{ u} = \frac{1}{12} \times \frac{12 \times 10^{-3}}{6.02205 \times 10^{23}} \text{ kg} = 1.66056 \times 10^{-27} \text{ kg}. \tag{1.2}$$

The energy equivalent of this amount of mass is

$$\begin{aligned}
1\text{u} &= 1.66056 \times 10^{-27} \times c^2 \\
&= 1.66056 \times 10^{-27} \times 8.98755 \times 10^6 \\
&= 931.502 \text{ MeV}
\end{aligned} \tag{1.3}$$

If Δm is in gram, then E will be in MeV per gram atom.

The binding energy of a nucleus, when divided by the number of nucleons, gives the mean binding energy per nucleon.

The binding energy per nucleon is a measure of the stability of the nucleus. The greater the binding energy per nucleon, the more stable the nucleus.

It is noted that if the value of binding energy is negative then the product of the nucleus or nuclei will be less stable than the reactant nucleus or nuclei. If the binding energy is positive, then the product nucleus is more stable than the reactant nucleus.

1.3 Nuclear Reactions

The reactions in which the nucleus of an atom itself undergoes a spontaneous change or interacts with other nuclei of lighter particles resulting in new nuclei (and one or more lighter particles) are called nuclear reactions. Here, the neutron reactions are mainly categorized into scattering, capture and fission. The exchange of energy between a neutron and a nucleus are based on the scattering reactions. In scattering reactions, scattering is used to describe the movement of the neutron after interaction. Scattering may be either elastic or inelastic. In elastic scattering, the exchange of energy between the neutron and the nucleus is kinetic. But in inelastic scattering, the kinetic energy of the neutron is transferred to the nucleus as potential energy.

Neutrons lose much of their kinetic energy due to the scattering collision with various nuclei in the medium through which the neutrons move and become slow with energies of an electron volt or less. Finally, the kinetic energy may be reduced to such an extent that the average is the same as that of the atoms (or molecules) of the medium. Since the value of the kinetic energy depends on the temperature, it is called thermal energy. Hence, the neutrons whose energies have been reduced to this extent are called thermal neutrons.

In inelastic scattering, the exited compound nucleus can emit its excess energy as gamma (γ) radiation and this process is known as radiative capture or capture. For example, ^{238}U is the most abundant naturally occurring isotope that has the following radiative capture:

$$^{238}_{92}U + ^{1}_{0}n \rightarrow ^{239}_{92}U + \gamma. \tag{1.4}$$

Here, the resulting nucleus ^{239}U is radioactive and decays with the emission of a negative beta particle.

Finally, the third type of interaction between neutrons and nuclei is fission, or it is called nuclear fission. This is an essential process in nuclear reactors, which will be discussed in the subsequent sections.

1.4 Nuclear Fusion

The process of combining lighter elements into a stable heavier nucleus is known as *nuclear fusion*. However, such processes can take place at reasonable rates only at very high temperatures of the order of several million degrees, which exist only in the interior of stars.

Such processes are therefore called thermonuclear reactions. Once a fusion reaction is initiated, the energy released in the process is sufficient to maintain the temperature and to keep the process going on.

For example, the hydrogen bomb is based on the fusion of hydrogen nuclei into heavier ones by the thermonuclear reactions with the release of enormous energy.

$$4_1\mathrm{H}^1 \rightarrow {}_2\mathrm{He}^4 + 2_{+1}e^0 + \mathrm{Energy} \tag{1.5}$$

1.5 Radioactivity

There are another class of nuclei that are unstable. They disintegrate (break up) spontaneously with the emission of electromagnetic radiation of very high energy. As such, the process of spontaneous transformation of a nucleus is known as radioactivity. Some of the unstable high atomic weight elements are radium, thorium and uranium. The emission or decay is associated with the radiation from the atomic nucleus, which is either an alpha particle (helium nucleus) or a beta particle (an electron). In many cases, gamma radiation comes out with the particle emission. The process of disintegration continues and after a number of stages of radiation, a stable nucleus is formed.

1.5.1 Rate of Radioactive Decay

The nuclei of a given radioactive species have a definite probability of decaying in unit time; this decay probability has a constant value characteristic of the particular nuclide. It remains the same, irrespective of the chemical or physical state of the element at all readily accessible temperatures and pressures. In a given specimen, the rate of decay at any instant is directly proportional to the number of radioactive atoms of the nuclide under considerations present at that instant. When N is the number of the particular radioactive atoms present at time t, then the decay rate can be represented as

$$\frac{dN}{dt} = -\lambda t \tag{1.6}$$

where λ is the decay constant of the radioactive nuclide.

If N_0 is the number of the particular radioactive atoms present at time zero and a time t, later, when N of these nuclei remain, Equation 1.6 is written as

$$\ln \frac{N}{N_0} = -\lambda t. \tag{1.7}$$

Equation 1.7 may be changed and encrypted as follows:

$$N = N_0 e^{-\lambda t}. \tag{1.8}$$

The ratio between the number of atoms disintegrating in unit time to the total number of atoms present at that time is called the decay constant of that nuclide.

In 1904, Rutherford introduced a constant known as the half-life period of the radioactive element for evaluating its radioactivity or for comparing its radioactivity with the activities of other radioactive elements. The half-life period of a radioactive element is defined as the time required by a given amount of the element to decay to one half of its initial value.

If $N = (N_0/2)$, then Equation 1.7 will be

$$\ln \frac{1}{2} = -\lambda t_{1/2} \Rightarrow t_{1/2} = \frac{\ln 2}{\lambda} = \frac{0.693}{\lambda}. \tag{1.9}$$

It has been observed that the half-life is inversely proportional to the decay constant.

The half-life period is a measure of the radioactivity of the element. The shorter the half-life period of an element, the greater is the number of the disintegrating atoms and the greater is its radioactivity (Sharma et al. 2008). The half-life period of different radioactive elements vary widely ranging from a fraction of a second to millions of years.

Since the total decay period of any element is not fixed, the total decay period of the radioactive element may be meaningless. As such, the average life is the ratio of sum of lives of the nuclei and total number of nuclei. In other words, the average life of an element is the inverse of its decay constant which is given as follows:

$$t_{av} = \frac{1}{\lambda}. \tag{1.10}$$

Substituting the value of λ in Equation 1.9, we get

$$t_{av} = \frac{t_{1/2}}{0.693} = 1.443 t_{1/2}. \tag{1.11}$$

The standard unit in radioactivity is curie (c), which is defined as that amount of any radioactive material, which gives 3.7×10^{10} disintegrations per second. In the SI system, the unit of radioactivity is Becquerel (Bq).

1.5.2 Radioactive Equilibrium

Let us consider a disintegration series in which A, B, C, D, etc. are some of the intermediate consecutive atoms (i.e. between the parent element and the final stable isotope)

$$\cdots A \rightarrow B \rightarrow C \rightarrow D \cdots$$

A stage may come when the amounts of A, B, C, D, etc. become constant, which is so because their rates of disintegration become equal. Further, if N_A, N_B, N_C and N_D, etc. represent the number of atoms of A, B, C, D, etc. at equilibrium, then

$$-\frac{dN_A}{dt} = -\frac{dN_B}{dt} = -\frac{dN_C}{dt} = \cdots \tag{1.12}$$

But

$$-\frac{dN_A}{dt} = k_A N_A, \quad -\frac{dN_B}{dt} = k_B N_B \quad \text{and} \quad -\frac{dN_C}{dt} = k_C N_C.$$

Equation 1.12 becomes

$$k_A N_A = k_B N_B = k_C N_C = \cdots \tag{1.13}$$

$$\frac{N_A}{N_B} = \frac{k_B}{k_A} = \frac{(t_{1/2})_A}{(t_{1/2})_B} \tag{1.14}$$

Thus, the amounts present at equilibrium are inversely proportional to their disintegration constants or directly proportional to their half-lives.

1.5.3 Radioactive Disintegration Series

A new element is formed when a radioactive element disintegrates and emits α or β particle. The element which emits α or β particle is called the parent element and the new element formed is called the daughter element. If the daughter element is radioactive, it again disintegrates by emitting α or β particle forming a new element. Thus, the daughter element becomes a parent element and a new daughter element is produced. This process of disintegration goes on till the end product is a stable isotope.

It may be noticed that in these series, sometimes a branched disintegration takes place (the same radioactive element can disintegrate in two different ways forming two completely different radioactive isotopes). However, both the radioactive isotopes thus formed disintegrate further to form the same radioactive isotope.

1.5.4 Artificial Radioactivity

During the artificial transmutation of elements, it is observed that quite often the product obtained is also radioactive, though its half-life period is usually very small. This phenomenon in which the artificial disintegration of a stable nucleus leads to the formation of a radioactive isotope is called artificial radioactivity.

It may be mentioned that as the radioactive isotopes formed by the artificial transmutation have usually very short half-life periods, they are very rare in nature. This is because as soon as they are formed, they decay. In some cases, instead of positrons, electrons (β particle) are emitted by the artificial radioactive isotopes produced.

Natural Radioactivity	Artificial Radioactivity
It involves spontaneous disintegration of unstable nuclei with emission of α or a β particles or γ radiations giving rise to new nucleus.	Here, stable nuclei are bombarded with high-energy particles to produce radioactive elements.
It cannot be controlled.	It can be controlled by controlling the speed of the bombarding particles.
It is shown by the elements with high atomic number and mass number.	It can be induced even in the lighter elements.

1.6 Nuclear Fission

When a nucleus is bombarded with some subatomic particles (like α particles, neutrons, protons, etc.), then these particles are either captured by the nucleus or the nucleus disintegrates ejecting some other subatomic particles. So, the new element formed has a mass either slightly greater than or slightly smaller than that of the original element.

The splitting of a heavier atom into a number of fragments of much smaller mass by suitable bombardment with subatomic particles, with the liberation of a huge amount of energy, is called *nuclear fission*.

When uranium-235 was hit by slow neutrons, it was split up into a number of fragments, each of mass much smaller than that of uranium. The two fragments were barium and krypton (Glasstone and Sesonke 2004).

$$_{92}U^{235} + _0n^1 \rightarrow _{56}Ba^{140} + _{36}Kr^{93} + 3\,_0n^1 + \text{Energy} \qquad (1.15)$$

The neutron released from the fission of the first uranium atom can hit three other uranium atoms, each of which again releases three neutrons, each of which can further hit one uranium atom and so on. In this way, a chain reaction is set up and results into the liberation of a huge amount of energy. Another aspect which is extremely important for a chain reaction to continue is that the fissionable material must have a minimum size. If the size is smaller than the minimum size, the neutrons escape from the sample without hitting the nucleus, causing fission, and thus the chain reaction stops.

The minimum mass that the fissionable material must have so that one of the neutrons released in every fission hits another nucleus and causes fission (so that the chain reaction continues at a constant rate) is called the critical mass. If the mass is less than that of the critical mass, it is called sub-critical. In this case, many neutrons released in fission are able to hit the other nuclei and thus the numbers of fissions multiply in the chain reaction. The shape and the density of the packing of the material are also significant for the nuclear fission.

Nuclear Fission	Nuclear Fusion
It involves breaking up of a heavier nucleus into lighter nuclei.	It involves the union of two or more lighter nuclei to from a heavier nucleus.
It is a chain process.	It is not a chain process.
It is initiated by neutrons of suitable energy and does not need high temperature.	It is initiated by very high temperatures.
It can be controlled and the energy released can be harnessed for useful purposes.	It is difficult to control this process.
A larger number of radioactive isotopes are formed and there is nuclear waste.	There is no nuclear waste in this process.
It requires a minimum size fissionable material, and if the size of the material exceeds the critical size, the reaction becomes explosive.	There is no limit to the size of the fuel for the reaction to start. However, the fuel does not undergo fusion until heated to a very high temperature (a few million degrees).

1.7 Principles, Production, and Interaction of Neutrons with Matter

Nuclear reactions and the behaviour of subatomic particles are basic to core design. Here, the main purpose is to provide details at the review level. We treat such processes as radioactivity decay, neutron scattering, radiative capture, and fission. In general, these processes are associated with the emission of subatomic particles and radiations as well as their interaction with matter. We use the general term 'radiation' to include both material particles and true electromagnetic radiation.

1.7.1 Production of Neutrons

Alpha (α) particles are highly energetic positively charged particles, which are emitted by radioactive substances, and the amount of their positive electricity is equal to two units of electronic charge. The mass of the α particle is equal to that of a helium atom. They are emitted during the radioactive disintegration of certain heavy elements like uranium or radium, etc. When mono-energetic alpha particles emitted from a radioactive substance are allowed to pass through a very thin metal foil, they are found to be scattered in different directions with respect to the direction of the collimated beam of the incident particles. Though, by far, a great majority of the particles are scattered at small angles (greater than 90°).

Such large-angle scattering cannot be explained on the basis of the Thomson model of the atom. The electrostatic repulsive force between the positive charge of the scattering atom and α-particle depends inversely upon the square of the distance (q) of the α-particle from the centre of the charge of the latter and directly upon the portion of the positive charge of the scattering atom contained within the sphere of the radius q. The angle of scattering increases with the increasing effective charge of the atom repelling the α-particle and decreases with the increasing distance from the centre of the atom at which the α-particle passes by the atom. When the α-particle passes by at a relatively large distance from the centre of the atom ($\sim 10^{-10}$ m near the periphery of the atom), this distance becomes so large that the angle of scattering is quite small, even though the entire positive charge of the atom repels it.

On the other hand, when the α-particle passes by at a relatively close distance from the centre of the atom, the effective charge repelling it is so small that the angle of scattering is again quite small.

In 1911, Rutherford proposed a new model of the atom. According to him, the entire positive charge of an atom and almost its entire mass are contained within a very small sphere near the centre. This positively charged core is known as the atomic nucleus. When an α-particle passes by very close to the centre of the atom, it feels a strong electrostatic repulsive force due to the entire positive charge of the atom and hence is scattered at a relatively large angle.

However, the Rutherford model of the atom has one serious drawback. As such, the electromagnetic theory of light predicts that the revolving electrons (due to their centripetal acceleration) should continually emit electromagnetic radiation so that they move spirally inwards and ultimately plunge into the nucleus. In 1913, Niels Bohr suggested a way out of the difficulty, which however involve entirely new concepts and that were at variance with some of the fundamental concepts of classical mechanics and Maxwell's electromagnetic theory of light.

This is known as Bohr's quantum theory. The quantum theory, in a more developed form at present, constitutes the theoretical basis of all subatomic phenomena.

In 1920, James Chadwick has determined the nuclear charge of several elements on the basis of Rutherford's theory of the α-particle scattering. To increase the number of scattered α-particles, Chadwick used a narrow ring-shaped scattering foil mounted on a suitable frame for the experiment. Since the atom as a whole is electrically neutral, the nuclear charge should be equal to the atomic number. Chadwick's determination of nuclear charge for the different elements confirmed this most conclusively.

1.7.2 Neutron Reactions and Radiation

The nature of the radiation emitted from a radioactive substance was investigated by Rutherford in 1904, by applying electric and magnetic fields to the radiations. It is observed that on applying the field, the rays emitted from the radioactive substances are separated into three types, called α, β and γ rays. The α rays are deflected in a direction that shows that they carry a positive charge; the β rays are deflected in the opposite direction, showing that they carry a negative charge and γ rays are not deflected at all, showing that they carry no charge. Furthermore, β rays are deflected to a much greater extent than α rays, which indicates that particles in the β rays are much lighter than those in the α rays. The properties of each of these rays have been studied in detail and briefly described next (Barik et al. 2011).

1.7.2.1 Properties of α Ray

- The direction of deflection of the α rays in the electric and magnetic fields shows that they carry a positive charge. The charge and mass of the particles present in the α rays (called α-particles) have been investigated by suitable experiments. It is found that each α particle carries two units of positive charge and has mass nearly four times that of a hydrogen atom. Thus, the α-particles are like the helium nuclei.
- The velocity of α rays is found to be nearly 1/10th to 1/20th of that of light, depending upon the nature of source.
- Alpha (α) rays ionize the gas through which they pass. The ionization takes place due to the knockout of the electrons from the molecules of the gas when the high-speed heavy α-particles hit these molecules.
- Since α-particles are heavy, they cannot pass through thick sheets of metal. In other words, α rays have a low penetrating power. It has been observed that they can penetrate through air only to a distance of about 7 cm and then they are absorbed in the air. Similarly, they can be stopped by an aluminium foil less than 1/10th of a mm in thickness.
- Alpha (α) rays affect a photographic plate and produce luminescence when they strike a zinc sulphide screen.

1.7.2.2 Properties of β Ray

- The direction of deflection of β rays in the electric and magnetic fields shows that they carry negative charge. The charge and mass of the particles present in these rays have been determined, and it is found that these particles possess the same charge and mass as that of the electrons. Thus, β-particles are nothing but electrons.
- As in case of α rays, the velocity of the β rays depends upon the nature of the source. However, being much lighter than the α-particles, β-particles travel at a higher speed than the α-particles. The speed of the β-particles varies from 3% to 99% of that of light, that is in some cases it approaches the velocity of light.
- In spite of a higher speed than α-particles, β-particles have less momentum or kinetic energy than the α-particles because of their smaller mass. Hence, β-particles have lower ionizing power than α-particles. The ionizing power of β-particles is about 1/100th of that of α-particles.
- Because of their smaller mass and higher speed, β-particles are much more penetrating than the α-particles. They can penetrate through an aluminium foil several

mm in thickness or through a lead sheet about 3 mm in thickness. Their penetrating power is about 100 times greater than that of α rays.

- Like α rays, β rays affect a photographic plate and the effect is much higher. However, there is no significant effect on a zinc sulphide screen because of their lower kinetic energy.

1.7.2.3 Properties of γ Ray

- They are not deflected in the electric and magnetic fields, showing thereby that they do not carry any charge. In fact, in every respect, these rays are found to be an electromagnetic radiation of the same nature as x-rays except that these are of a shorter wavelength than x-rays. These rays do not have any mass and hence cannot be considered to be made up of particles. It is interesting to mention here that the production of γ rays does not take place simultaneously with the α and β rays but it takes place subsequently. It is believed that when an α or β particle is emitted, the nucleus becomes exited, that is it has a higher energy and emits the excess energy in the form of radiation which form γ rays. In some cases, only α and β rays are emitted and γ rays are not present at all.
- They travel with the same velocity as that of light.
- As they do not have any mass, their ionizing power is very poor.
- They have the highest penetrating power out of all the three types of radiations. Their penetrating power is about 100 times more than that of β rays. Thus, they can penetrate through lead sheets as thick as 150 mm.
- Gamma (γ) rays have very little effect on the photographic plate or zinc sulphide screen.

1.7.3 Inelastic and Elastic Scattering of Neutrons

In the nuclear reactor, a neutron undergoes scattering and these are inelastic and elastic. When a fast neutron undergoes inelastic scattering, then it is first captured by the target (scattering) nucleus to form an excited (virtual) state of compound nucleus. A neutron of lower kinetic energy is then emitted and leaving the target nucleus in an excited (bound) state. In an inelastic scattering collision, some (or all) of the kinetic energy of the neutron is converted into the excitation (internal) energy of the target nucleus. This excess energy is subsequently emitted as one or more photons (a particle representing a quantum of light or other electromagnetic radiation) of gamma (γ) radiations, called inelastic scattering gamma (γ) rays. The total energy of these gamma (γ) rays is equal to the excess energy of the excited state of the target nucleus.

Let us consider E_1 to be the total kinetic energy of the neutron and the target nucleus before collision, and E_2 is the kinetic energy after collision. If E_γ is the total energy emitted as gamma radiation, then

$$E_1 = E_2 + E_\gamma. \tag{1.16}$$

It is evident that in inelastic scattering, kinetic energy is not conserved. Nevertheless, there is conservation of momentum, so if E_γ were known, the mechanics of the process could be solved. This is not necessary, however, for the present purpose. Since the kinetic energy of the target nucleus is (in general) negligible in comparison with that of the neutron, it follows that, in an inelastic collision, the initial energy of the neutron must exceed the minimum excitation energy of the target nucleus.

For elements of moderate and high mass number, the minimum excitation energy, that is the energy of the lowest excited state of the target nucleus above the ground state, is usually from 0.1 (or so) to 1 MeV. Hence, only neutrons with an energy exceeding this amount can be inelastically scattered as a result of nuclear excitation. With the decreasing mass number of the nucleus, there is a general tendency for the excitation energy to increase so that the neutrons have higher energies if they are to undergo inelastic scattering. The threshold energy for such scattering in oxygen is about 6 MeV, and in hydrogen the process does not occur at all. Exceptions to the foregoing generalizations are the magic nuclei – heavy nuclei of this type. For example, lead (82 protons) and bismuth (126 neutrons) behave like light nuclei with respect to inelastic scattering.

For elements of low atomic number, the excitation energy of the first (lowest) excited state is large. Hence, the total inelastic scattering gamma ray energy will be large (several MeV). However, for elements of moderate and high mass number, the energy will be lower, usually in the range of 0.1–1 MeV. Magic nuclei are exceptional in this respect, as indicated in the preceding paragraph.

Another type of inelastic scattering can occur at low neutron energies, generally below a few eV, if the scattering atom (or nucleus) is not free but is bound in a molecule or in a solid. Such a scattering system has discrete quantum states of internal energy associated with the vibrations of the atoms in a molecule (or in a solid) and the rotation of the molecule as a whole. Both vibrational and rotational motions can occur in polyatomic gases and liquids. But in a solid only the vibrational motion is significant.

In the collision of a low-energy neutron with a bound scattering atom, there can be loss or gain of internal energy associated with a change in the vibrational and rotational states. Such a collision would therefore be described as inelastic scattering. The energy of the scattered neutron may be greater or less than the energy prior to the collision. There is no compound nucleus formation in this low-energy type of inelastic scattering. However, for neutron energies above a few eV, sufficient energy is transferred to the scattering atom to permit it to be free or behave as if it were free. In this case, inelastic scattering, accompanied by changes in vibrations and rotational energies, does not occur.

1.7.3.1 Elastic Scattering

As discussed, inelastic scattering of neutrons is limited to certain (high or low) energy ranges, but in elastic scattering collisions, there are no restrictions on the exchange of energy between the neutron and the target nucleus. Hence, elastic scattering is positive with all nuclei, free or bound, and neutrons of all energies. In elastic scattering, there is no change in the internal energy of the scattering system and kinetic energy of the neutron exceeds that of the scattering nucleus, some of the kinetic energy of the former may be transferred to the latter and vice versa.

Elastic collisions of neutrons with nuclei are of two types. One of the major interesting nuclear reactor systems is called potential scattering; for nuclei of low mass number, scattering occurs with neutrons that have energies up to a few MeV. In this type of scattering, there is no compound nucleus formation and scattering results from short-range forces acting on the neutron as it approaches the nucleus. In the other type of elastic scattering, known as resonance (or compound nucleus) scattering, a compound nucleus is formed when the scattering nucleus absorbs the neutron; the compound nucleus then expels a neutron leaving the target nucleus in its ground state but usually with a different (larger) kinetic energy. As a general rule in situations of present interest, elastic scattering can be treated as a 'billiard ball' type of collision. The behaviour can then be analyzed by means of the conservation of kinetic energy and of momentum.

After a sufficient number of scattering collisions, the speed of a neutron is reduced to such an extent that it has approximately the same average kinetic energy as the atoms (or molecules) of the scattering medium. The energy depends on the temperature of the medium, and so it is called thermal energy. Thermal neutrons are neutrons which are in thermal equilibrium with the atoms (or molecules) of the medium in which they are present. A particular thermal neutron undergoes collisions with the nuclei of the ambient medium and may gain or lose energy in a non-absorbing medium. There is no net energy change for all the neutrons.

1.7.4 Maxwell–Boltzmann Distribution

If a large number of neutrons are introduced into an infinite, non-absorbing scattering medium, that is a medium from which neutrons are neither lost by escape nor by absorption, a state of thermal equilibrium will be attained. In this state, the probabilities are equal (that a neutron will gain or lose energy in a collision with a scattering nucleus). The kinetic energies of the neutrons would then be represented by the Maxwell–Boltzmann distribution commonly referred to as the 'Maxwellian' distribution. Strictly speaking, such a distribution can be attained only if the scattering nuclei are not bound but are free to move. It will be seen shortly that this and other required conditions are not satisfied in an actual reactor system; nevertheless, it is useful to assume as a first approximation that the neutrons become thermalized to the extent that the Maxwellian distribution exists.

The Maxwell–Boltzmann distribution law can be derived from the kinetic theory of gases or by the methods of statistical mechanics. For the present purpose, the kinetic energy distribution of neutrons in thermal equilibrium at the absolute (kelvin) temperature T may be expressed by

$$\frac{dn}{n} = \frac{2\pi}{\left(\pi kT\right)^{3/2}} e^{-E/kT} E^{1/2} dE \qquad (1.17)$$

where
  dn is the number of neutrons with energies in the range from E to $E+dE$
  n is the total number of neutrons in the system
  k is the Boltzmann constant

The equation may be written in a slightly different form by letting $n(E)$ represent the number of neutrons of energy E per unit energy interval. Then $n(E)dE$ is the number if neutrons having energies in the range from E to $E+dE$, which is equivalent to dn in Equation 1.17. The latter may thus be written as

$$\frac{n(E)}{n} = \frac{2\pi}{\left(\pi kT\right)^{3/2}} e^{-E/kT} E^{1/2} \qquad (1.18)$$

where the left side represents the fraction of the neutrons having energies within a unit energy interval at energy E. The right side of Equation 1.18 can be evaluated for various E at a given temperature and the Maxwellian distribution curve is obtained in this manner, indicating the variation of $n(E)/n$ with the kinetic energy E of the neutrons.

Bibliography

Azekura, K. 1980. New finite element solution technique for neutron diffusion equations. *Journal of Nuclear Science and Technology* 17(2):89–97.

Barik, N., Das, L. K. and Shrma, K. N. 2011. *A Textbook of +2 Physics*, Vols. I and II. Kalyani Publisher, New Delhi, India.

Ghoshal, S. N. 2010. *Nuclear Physics*. S. Chand & Co. Ltd.

Glasstone, S. and Sesonke, A. 2004. *Nuclear Reactor Engineering*. CBS Publishers and Distributors Private Limited.

Hetrick, D. L. 1971. *Dynamics of Nuclear Reactors*. University of Chicago Press.

Sharma, Y. R., Nanda, R. N. and Das, A. K. 2008. *A Textbook of Modern Chemistry*, 14th edn. Kalyani Publisher, New Delhi, India.

Stacey, W. M. 2007. *Nuclear Reactor Physics*. Wiley-VCH.

Wood, J. and de Oliveira, C. 1984. A multigroup finite element solution of the neutron transport equation – I. *Annals of Nuclear Energy* 11(5):229–243.

2

Neutron Diffusion Theory

Neutron diffusion theory is the backbone of a nuclear reactor. It gives scientific insights to study various nuclear design problems. The neutrons are characterized by a single energy or speed in one-group diffusion theory and the model allows preliminary design estimates (Keepen 1964). Multigroup diffusion and transport theory may also be analyzed by assuming the same mathematical models used for primary design estimates. Here, the scattering of neutron depends on the cross section of neutron reactions, rates of neutron reactions, fission neutrons, prompt neutrons and delayed neutrons.

2.1 Cross Section of Neutron Reactions

Neutron interactions with atomic nuclei can be made quantitative and explained by means of the concept of *cross sections*. The rate at which any particular nuclear reaction occurs to the action of neutrons for a given material is exposed depends upon the number of neutrons, their speed and nature of the nuclei in the specified material. The cross section of a target nucleus for any given reaction is a measure of the probability of a particular neutron–nucleus interaction. It is also a property of the nucleus and of the energy of the incident neutrons. In other words, the nuclear cross section σ for a particular reaction is defined as the average number of individual processes occurring per target nucleus per incident neutron.

The cross section σ for a particular process to a single nucleus is generally called the *microscopic cross section*. As the target material contains N nuclei per m², the quantity $N\sigma$ that is equivalent to the total cross section of the nuclei per m² is defined as the *macroscopic cross section* of the material for the process. Symbolically, *macroscopic cross section* is represented by Σ; it is therefore defined as (Foster and Wright 1978)

$$\Sigma = N\sigma \ \mathrm{m}^{-1}, \tag{2.1}$$

with dimensions of a reciprocal length.

2.1.1 Transport Cross Section

The use of the transport cross section in the effect of the scattering angular distribution of the motion of a neutron has the following assumptions:

1. An infinite medium.
2. A purely scattering medium without absorption.
3. The energy of the neutron unchanged as a result of a collision with the nuclei of the medium.
4. After each collision, the particle travels a scattering mean free path λ_s and is deflected by an angle θ.

After the introduction of the neutron to the system, the neutron moves a distance

$$\bar{Z}_0 = \lambda_s = \frac{1}{\Sigma_s} \tag{2.2}$$

where Σ_s is the macroscopic scattering cross section and has the dimension of a reciprocal length.

After the first collision along the z-axis, the projected distance travelled by the neutron is (as shown in Figure 2.1)

$$Z_1 = \lambda_s \cos \theta_1$$

The average value of Z_1 becomes

$$\bar{Z}_1 = \lambda_s \langle \cos \theta_1 \rangle = \lambda_s \bar{\mu} \tag{2.3}$$

where $\bar{\mu} = 2/(3A)$; A is the mass number of the nuclei in the scattering medium.

The distance travelled after the second collision (Figure 2.1) is

$$Z_2 = \lambda_s \langle \cos \theta_1 \rangle \langle \cos \theta_2 \rangle.$$

Now, the average value of Z_2 is

$$\bar{Z}_2 = \lambda_s \langle \cos \theta_1 \rangle \langle \cos \theta_2 \rangle \approx \lambda_s \langle \cos \theta_1 \rangle^2 = \lambda_s \bar{\mu}^2. \tag{2.4}$$

Generalizing the collisions and average values of the distance travelled by the neutrons for n collisions are

$$\bar{Z}_n = \lambda_s \bar{\mu}^n, \quad \text{for } n = 0,1,2,\dots \tag{2.5}$$

Since $\bar{Z}_n \to 0$ when $n \to \infty$, it implies that the neutron could be scattered in either direction with equal probability, at successive collisions. Hence, after successive collisions, the

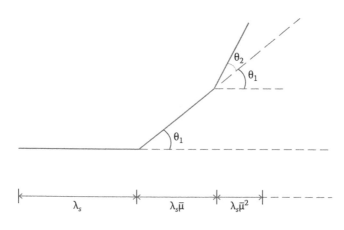

FIGURE 2.1
Transportation of a particle and its geometry.

neutron forgets its original direction of motion, which is characteristic of Markov chains. As such, the distance covered by the system of a transport mean free path becomes

$$\lambda_{tr} = \bar{Z}_0 + \bar{Z}_1 + \cdots = \sum_{n=0}^{\infty} \lambda_s \bar{\mu}^n$$
$$= \lambda_s \left(1 + \bar{\mu} + \bar{\mu}^2 + \cdots\right)$$
$$= \frac{\lambda_s}{1 - \bar{\mu}}, \quad \forall \bar{\mu} < 1 \tag{2.6}$$

The transport cross section is defined as

$$\Sigma_{tr} = \frac{1}{\lambda_{tr}} = \frac{1 - \bar{\mu}}{\lambda_s} = \Sigma_s \left(1 - \bar{\mu}\right) \tag{2.7}$$

If absorption is present, the definition of transport cross section will be

$$\Sigma_{tr} = \Sigma_a + \Sigma_s \left(1 - \bar{\mu}\right)$$
$$= \left(\Sigma_a + \Sigma_s\right) - \Sigma_s \bar{\mu} = \Sigma_t - \Sigma_s \bar{\mu} \tag{2.8}$$

where Σ_t is the total macroscopic cross section.

2.2 Rates of Neutron Reactions

Consider a mono-energetic neutron beam having a neutron density n (the number of neutrons per m³). If the neutron speed is v m/s, then nv is the number of neutrons falling on 1 m² of the target material per second. If σ m² is the effective area per single nucleus, for a given reaction (or reactions) and neutron energy, we have the effective area Σ of all the nuclei per m³ of the target. Hence, the product of effective area per single nucleus and the effective area, that is, Σnv give the number of interactions (between neutrons and nuclei) per m³ of target material per second.

$$\text{Rate of neutron interaction} = \Sigma nv \text{ interactions/m}^3 \cdot \text{s} \tag{2.9}$$

This is an important result as it gives the number of neutrons per second involved in any interaction (or interactions) with 1 m³ of material for which Σ is the macroscopic cross section. Sometimes, it may be written in a slightly different form by replacing the neutron flux in place of the neutron density. The neutron flux is defined as the product of the neutron density and the velocity, that is,

$$\phi = nv. \tag{2.10}$$

It is expected in units of neutrons/m²·s, which is equal to the total distance in metres travelled in 1 s by all neutrons present in 1 m³. Sometimes, it is referred to as the track length. Substituting ϕ for nv in Equation 2.9, it follows that

$$\text{Rate of neutron interaction} = \Sigma \phi \text{ interactions/m}^3 \cdot \text{s} \tag{2.11}$$

This flux is a special case of the angular flux, which adds a spatial angular dependency to the neutron density. The angular flux is a scalar. For the present discussion, we neglect the angular dependency by assuming that the integrated effect is negligible.

Fick's law of diffusion states: 'If the concentration of a solute in one region is greater than in another of a solution, the solute diffuses from the region of higher concentration to the region of lower concentration' (Lamarsh 1983). Using Fick's law, the diffusion approximation in reactor theory is discussed next.

Let us assume the following:

1. Consider an infinite medium.
2. The cross sections are constants, which are independent of positions and implying a uniform medium.
3. Scattering of neutrons are isotropic in the laboratory system.
4. The neutron flux function slowly varies with respect to the position.
5. We use a one-speed system where the neutron density is not a function of time.
6. We use a steady-state system where the neutron density is not a function of time.
7. There is no fission source in the system.

Some of these assumptions will be relaxed afterwards. For instance, the diffusing medium will be taken as finite in size rather than infinite.

We shall attempt to calculate the current density at the centre of the coordinate system. The vector $\underline{J}$ is given by

$$\underline{J} = J_x \hat{i} + J_y \hat{j} + J_z \hat{k} \tag{2.12}$$

so we must evaluate the components J_x, J_y, J_z.

These net current components can be written in terms of the partial axial currents as

$$\begin{aligned}
J_x &= J_x^+ - J_x^- \\
J_y &= J_y^+ - J_y^- \\
J_z &= J_z^+ - J_z^-
\end{aligned} \tag{2.13}$$

Let us investigate the estimation of one single component J_z crossing the element of area dS_z at the origin of the coordinate system in the negative z direction.

Due to the scattering collision of the neutrons, every neutron passes through the area dS_z in the x–y plane. Neutron scattering above x–y plane will thus flow downwards through the area dS_z.

Consider the volume element shown in Figure 2.2, which may be represented as

$$dV = r^2 \sin\theta\, dr\, d\theta\, d\varphi. \tag{2.14}$$

The number of scattering collisions occurring per unit time in the volume element dV is represented as

$$\Sigma_s \phi(r) dV = \Sigma_s \phi(r) r^2 \sin\theta\, dr\, d\theta\, d\varphi \tag{2.15}$$

where $\phi(r)$ and Σ_s are called the particle flux and macroscopic scattering cross section.

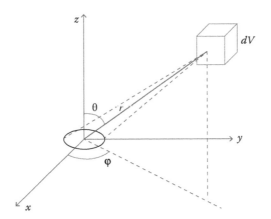

FIGURE 2.2
Geometry for the neutron current.

Since scattering is isotropic in LAB system (Glasstone and Sesonke 2004) and the fraction of scattering neutron arriving to dS_z is subtended by the solid angle $d\Omega$, which may be written as

$$\frac{d\Omega}{\Omega} = \frac{dS/r^2}{4\pi} = \frac{dS_z \cos\theta}{4\pi r^2} \qquad (2.16)$$

The number of neutrons scattered per unit time in dV reaching dS_z and the exponential factor $e^{-\Sigma_t r}$ which is attenuated in the medium are

$$dN = e^{-\Sigma_t r}\phi(r)\Sigma_s r^2 \sin\theta dr d\theta d\varphi \cdot \frac{dS_z \cos\theta}{4\pi r^2} \qquad (2.17)$$

The partial current J_z^- can now be written as

$$J_z^- = \frac{\int dN}{dS_z} = \frac{\Sigma_s}{4\pi}\int_0^\infty\int_0^{\pi/2}\int_0^{2\pi} e^{-\Sigma_t r}\phi(r)\sin\theta\cos\theta dr d\theta d\varphi. \qquad (2.18)$$

Since $\phi(r)$ is an unknown function, so we expand it in a Taylor series assuming that it varies slowly with position, which will be

$$\phi(r) = \phi_0 + x\left.\frac{\partial\phi}{\partial x}\right|_0 + y\left.\frac{\partial\phi}{\partial y}\right|_0 + z\left.\frac{\partial\phi}{\partial z}\right|_0 + \cdots \qquad (2.19)$$

Writing x, y and z in spherical coordinates, Equation 2.19 becomes

$$\phi(r) = \phi_0 + r\sin\theta\cos\varphi\left.\frac{\partial\phi}{\partial x}\right|_0 + r\sin\theta\sin\varphi\left.\frac{\partial\phi}{\partial y}\right|_0 + r\cos\theta\left.\frac{\partial\phi}{\partial z}\right|_0 + \cdots \qquad (2.20)$$

Substituting Equation 2.20 into Equation 2.18, we have

$$J_z^- = \frac{\Sigma_s}{4\pi} \int_0^\infty \int_0^{\pi/2} \int_0^{2\pi} e^{-\Sigma_t r} \left[\phi_0 + r\sin\theta\cos\varphi \frac{\partial\phi}{\partial x}\bigg|_0 + r\sin\theta\sin\varphi \frac{\partial\phi}{\partial y}\bigg|_0 + r\cos\theta \frac{\partial\phi}{\partial z}\bigg|_0 \right] \sin\theta\cos\theta \, dr \, d\theta \, d\varphi.$$

(2.21)

Integrating the terms containing $\cos\varphi$ and $\sin\varphi$ to zero over the interval $\varphi \in [0, 2\pi]$, we get

$$J_z^- = \frac{\Sigma_s}{4\pi} \int_0^\infty \int_0^{\pi/2} \int_0^{2\pi} e^{-\Sigma_t r} \left[\phi_0 + r\cos\theta \frac{\partial\phi}{\partial z}\bigg|_0 \right] \sin\theta\cos\theta \, dr \, d\theta \, d\varphi.$$

(2.22)

The first term may be evaluated as follows:

$$\begin{aligned}
I_1 &= \frac{\Sigma_s}{4\pi} \phi_0 \int_0^\infty \int_0^{\pi/2} \int_0^{2\pi} e^{-\Sigma_t r} \sin\theta\cos\theta \, dr \, d\theta \, d\varphi \\
&= \frac{\Sigma_s}{4\pi} \phi_0 \cdot 2\pi \cdot \frac{1}{\Sigma_t} \cdot \frac{1}{2} \\
&= \frac{1}{4} \frac{\Sigma_s}{\Sigma_t} \varphi_0
\end{aligned}$$

The second term becomes

$$\begin{aligned}
I_2 &= \frac{\Sigma_s}{4\pi} \frac{\partial\phi}{\partial z}\bigg|_0 \int_0^\infty \int_0^{\pi/2} \int_0^{2\pi} e^{-\Sigma_t r} \sin\theta\cos^2\theta \, dr \, d\theta \, d\varphi \\
&= \frac{\Sigma_s}{4\pi} \frac{\partial\phi}{\partial z}\bigg|_0 \cdot 2\pi \cdot \frac{1}{\Sigma_t^2} \cdot \frac{1}{3} \\
&= \frac{1}{6} \frac{\Sigma_s}{\Sigma_t^2} \frac{\partial\phi}{\partial z}\bigg|_0
\end{aligned}$$

Thus, Equation 2.22 transforms to

$$J_z^- = \frac{1}{4} \frac{\Sigma_s}{\Sigma_t} \varphi_0 + \frac{1}{6} \frac{\Sigma_s}{\Sigma_t^2} \frac{\partial\phi}{\partial z}\bigg|_0.$$

(2.23)

Similarly,

$$J_z^- = \frac{1}{4} \frac{\Sigma_s}{\Sigma_s} \varphi_0 - \frac{1}{6} \frac{\Sigma_s}{\Sigma_t^2} \frac{\partial\phi}{\partial z}\bigg|_0.$$

(2.24)

Substituting Equations 2.23 and 2.24 into Equation 2.13, we get

$$\begin{aligned}
J_x &= -\frac{1}{3} \frac{\Sigma_s}{\Sigma_t^2} \frac{\partial\phi}{\partial x}\bigg|_0 \\
J_y &= -\frac{1}{3} \frac{\Sigma_s}{\Sigma_t^2} \frac{\partial\phi}{\partial y}\bigg|_0 \\
J_z &= -\frac{1}{3} \frac{\Sigma_s}{\Sigma_t^2} \frac{\partial\phi}{\partial z}\bigg|_0
\end{aligned}$$

(2.25)

Substituting Equation 2.12, we get the expression for the current density after dropping the evaluation at the origin notation. Since the origin of the coordinates is arbitrary, we have

$$
\begin{aligned}
\underline{J} &= -\frac{1}{3}\frac{\Sigma_s}{\Sigma_t^2}\left(\frac{\partial\phi}{\partial x}\hat{i} + \frac{\partial\phi}{\partial y}\hat{j} + \frac{\partial\phi}{\partial z}\hat{k}\right) \\
&= -\frac{1}{3}\frac{\Sigma_s}{\Sigma_t^2}\nabla\phi \\
&= -D\nabla\phi
\end{aligned}
\tag{2.26}
$$

where
   ∇ is the gradient
   $D = (1/3)\left(\Sigma_s/\Sigma_t^2\right)$ is the diffusion coefficient

Thus, Fick's law for neutron diffusion is given by

$$
\underline{J} = -D\nabla\phi
\tag{2.27}
$$

It states that the current density vector is proportional to the negative gradient of the flux and establishes a relationship between them under the enunciated assumptions. Notice that the gradient operator turns the neutron flux, which is a scalar quantity into the neutron current, which is a vector quantity.

2.2.1 Neutron Diffusion Equation

To derive the neutron diffusion equation, we have to consider the following assumptions:

1. Use a one-speed or one-group approximation where the neutrons can be characterized by a single average kinetic energy.
2. Characterize the neutron distribution in the reactor by the particle density $n(r, t)$, which is the number of neutrons per unit volume at a position r' at time t. Its relationship to the flux is

$$
\phi(r', t) = vn(r', t)
\tag{2.28}
$$

We consider an arbitrary volume V and write the balance equation as follows.
   Time of change of the number of neutrons in V = Production rate in V – Absorptions in V – Net leakage from the surface of V.
   The first term is expressed mathematically as

$$
\frac{d}{dt}\left[\int_V n(r', t)dV\right] = \frac{d}{dt}\left[\int_V \frac{1}{v}\phi(r', t)dV\right] = \frac{1}{v}\left[\int_V \frac{\partial}{\partial t}\phi(r', t)dV\right]
$$

The production rate can be written as

$$
\int_V S(r', t)dV
$$

The absorption term is

$$\int_V \Sigma_a(r')\phi(r',t)dV$$

and the leakage term is

$$\int_S \underline{J}(r',t)\cdot\hat{n}ds = \int_V \nabla\cdot\underline{J}(r',t)dV$$

where we convert the surface integral to a volume integral by using Gauss theorem or the divergence theorem.

Substituting for the different terms in the balance equation, we get

$$\int_V \frac{1}{v}\frac{\partial\phi(r',t)}{\partial t}dV = \int_V S(r',t)dV - \int_V \Sigma_a(r')\phi(r',t)dV - \int_V \nabla\cdot\underline{J}(r',t)dV$$

Or,

$$\int_V \left(\frac{1}{v}\frac{\partial\phi(r',t)}{\partial t} - S(r',t) + \Sigma_a(r')\phi(r',t) + \nabla\cdot\underline{J}(r',t)\right)dV = 0 \qquad (2.29)$$

Since the volume V is arbitrary, we may write Equation 2.29 as

$$\frac{1}{v}\frac{\partial\phi(r',t)}{\partial t} = -\nabla\cdot\underline{J}(r',t) - \Sigma_a(r')\phi(r',t) + S(r',t)$$

We now use the relationship between $\underline{J}$ and ϕ (Fick's law) to write the diffusion equation

$$\frac{1}{v}\frac{\partial\phi(r',t)}{\partial t} = \nabla\cdot\left[D(r')\nabla\phi(r',t)\right] - \Sigma_a(r')\phi(r',t) + S(r',t) \qquad (2.30)$$

This is the basis of the development in reactor theory using diffusion theory.

2.2.2 Helmholtz Equation

The diffusion equation (2.30) is a partial differential equation of the parabolic type. It also describes the physical phenomena in heat conduction, gas diffusion and material diffusion.

This equation can be simplified in the case the medium is uniform or homogeneous such that D and Σ_a do not depend on the position r' as (Hetrick 1971)

$$S(r',t) = \frac{1}{v}\frac{\partial\phi(r',t)}{\partial t} - D\nabla^2\phi(r',t) + \Sigma_a\phi(r',t) \qquad (2.31)$$

where we used the fact that the divergence of the gradient leads to the Laplacian operator:

$$\nabla\cdot\nabla = \nabla^2$$

The Laplacian operator ∇^2 depends on the coordinates system used:

$$\text{Cartesian: } \nabla^2 \equiv \frac{\partial^2}{\partial x^2} + \frac{\partial^2}{\partial y^2} + \frac{\partial^2}{\partial z^2}$$

$$\text{Cylindrical: } \nabla^2 \equiv \frac{1}{r}\frac{\partial}{\partial r}\left(r\frac{\partial}{\partial r}\right) + \frac{1}{r^2}\frac{\partial^2}{\partial \theta^2} + \frac{\partial^2}{\partial z^2}$$

$$\text{Spherical: } \nabla^2 \equiv \frac{1}{r^2}\frac{\partial}{\partial r}\left(r^2\frac{\partial}{\partial r}\right) + \frac{1}{r^2 \sin\theta}\frac{\partial}{\partial \theta}\left(\sin\theta\frac{\partial}{\partial \theta}\right) + \frac{1}{r^2 \sin^2\theta}\frac{\partial^2}{\partial \varphi^2}$$

When the flux is not a function of time, we use the steady-state diffusion equation or the scalar Helmholtz equation

$$D\nabla^2\phi - \Sigma_a\phi + S = 0 \tag{2.32}$$

which is a partial differential equation of the elliptic type.
The Helmholtz equation can be written as

$$\nabla^2\phi(r') - \frac{1}{L^2}\phi(r') = -\frac{S(r')}{D} \tag{2.33}$$

where $L^2 = D/\Sigma_a$ is the diffusion length.

Bibliography

Azekura, K. 1980. New finite element solution technique for neutron diffusion equations. *Journal of Nuclear Science and Technology* 17(2):89–97.

Chakraverty, S. and Nayak, S. 2013a. Fuzzy finite element method in diffusion problems. In: *Mathematics of Uncertainty Modelling in the Analysis of Engineering and Science Problems*. IGI Global, pp. 309–328.

Chakraverty, S. and Nayak, S. 2013b. Non probabilistic solution of uncertain neutron diffusion equation for imprecisely defined homogeneous bare reactor. *Annals of Nuclear Energy* 62:251–259.

Duderstadt, J. J. and Hamilton, L. J. 1976. *Nuclear Reactor Analysis*. John Wiley & Sons, Inc., New York.

Foster, A. R. and Wright, R. L. 1978. *Basic Nuclear Engineering*. Allyn & Bacon.

Glasstone, S. and Sesonke, A. 2004. *Nuclear Reactor Engineering*. CBS Publishers and Distributors Private Limited.

Hetrick, D. L. 1971. *Dynamics of Nuclear Reactors*. University of Chicago Press.

Hetrick, D. L. 1993. *Dynamics of Nuclear Reactors*. American Nuclear Society, La Grange Park, IL.

Keepen, G. R. 1964. *Physics of Nuclear Kinetics*. Addison-Wesley.

Lamarsh, J. R. 1983. *Introduction to Nuclear Engineering*. Addison-Wesley Publishing Company.

Nayak, S. and Chakraverty, S. 2013. Non-probabilistic approach to investigate uncertain conjugate heat transfer in an imprecisely defined plate. *International Journal of Heat and Mass Transfer* 67:445–454.

Ragheb, M. 1982. *Lecture Notes on Fission Reactors Design Theory*. FSL-33. University of Illinois.

3

Fundamentals of Uncertainty

Uncertainty is an important aspect in experiment as well as modelling in various science and engineering problems. It has become one of the popular paradigmatic changes in recent decades. These changes have been demonstrated by a gradual transition from the traditional outlook, which suggested that uncertainties were undesirable in science and should be avoided by all possible means. Traditionally, science strived for certainty in all its manifestations (precision, specificity, sharpness, consistency, etc.); thus, uncertainties (impreciseness, non-specificity, vagueness, inconsistency, etc.) may have been regarded earlier as unscientific. In the modern view, uncertainties have been considered as an essential part of science, and it is the backbone of various real-world problems.

Generally, we deal with systems that are constructed as models of either some facet of reality or some desirable artificial objects. The need for constructing models is to understand some phenomenon of reality, be it natural or artificial, making decent predictions, learning to control the phenomenon in any desirable way and utilizing all these capabilities for various ends; models of the latter type are constructed for the purpose of prescribing operations by which a conceived artificial object can be constructed in such a way that desirable objective criteria are satisfied within given constraints.

As such, uncertainty plays an important role in various branches of engineering and science. These uncertainties may be quantified by two approaches, viz. parametric and non-parametric. In the parametric approach, uncertainties may be associated with system parameters, such as Young's modulus, mass density, Poisson's ratio, damping coefficient and geometric parameters, for structural dynamic problems. Various authors have quantified these parameters using statistical methods, for example the stochastic finite element method. On the other hand, uncertainties that occur due to incomplete data or information, impreciseness, vagueness, experimental error and different operating conditions influenced by the system may be handled by non-parametric approaches.

In recent years, the probabilistic approach has been adopted to handle the uncertainties involved in the systems. Probability distributions have been used in traditional deterministic model parameters that account for their uncertainty. Different frameworks have been proposed to quantify the uncertainties caused by randomness (aleatory uncertainty) as well as lack of knowledge (epistemic uncertainty), along with probability boxes (p-boxes) (Ferson and Ginzburg 1996), Bayesian hierarchical models (Gelman 2006), Dempter–Shafer's evidence theory (Dempster 1967; Shafer 1976) and parametric p-boxes using sparse polynomial chaos expansions (Schöbi and Sudret 2015a). These frameworks are generally referred to as imprecise probabilities. The uncertainty propagation of imprecise probabilities leads to an imprecise response. Input uncertainties are characterized and then propagated through a computational model. To reduce the computational effort, a well-known meta-modelling technique such as polynomial chaos expansions (Ghanem and Spanos 2003; Sudret 2014; Schöbi and Sudret 2015b) can be used. An algorithm has been proposed by Schöbi and Sudret (2015b) for solving imprecise structural reliability problems. This algorithm transforms an imprecise problem into two precise structural reliability problems, which reveals the possibilities for using traditional structural reliability analyses techniques.

Although the uncertainties are handled by various authors using probability density functions or statistical methods, these methods need plenty of data and also may not consider the vague or imprecise parameters. Accordingly, one may use the interval/fuzzy computation in the analysis of the problems. In this context, a few authors have used finite element method when the uncertain parameters are in terms of interval/fuzzy and it is called the interval/fuzzy finite element method (I/FFEM). Accordingly, a new computation method (Chakraverty and Nayak 2013; Nayak and Chakraverty 2013) with interval/fuzzy values was developed for reducing the computational effort. Applying the I/FFEM, we get either interval/fuzzy system of equations or eigenvalue problems. The solutions for the interval/fuzzy system of linear equations are studied by various researchers. A few authors have also discussed the method of the uncertain bound of eigenvalues. Sevastjanov and Dymova (2009) investigated a new method for solving both the interval/fuzzy equations for linear case. Friedman et al. (1998) used the embedding approach to solve the $n \times n$ fuzzy linear system of equations. Some authors (Abbasbandy and Alvi 2005; Allahviranloo et al. 2008; Li et al. 2010; Senthilkumar and Rajendran 2011) proposed various other methods for finding the uncertain solutions of fuzzy system of linear equations. They have considered the coefficient matrix as crisp. This literature shows that the uncertain systems may also be handled by using the non-probabilistic approach. Accordingly, both probabilistic and non-probabilistic approaches may be considered for investigating uncertain systems.

Both approaches, viz. probabilistic and non-probabilistic, have become popular among researchers in the field. They are discussed in the detail next.

3.1 Probabilistic Uncertainty

An experiment is a process of measurement or observation. The aim is to study the randomness in the experiment. A trial is a single performance of an experiment. Its result is called an outcome or a sample point. The set of all possible outcomes is called sample space S.

Let S be a finite sample space and E be an event in S, then the probability $P(E)$ of the event E is a real number assigned to E (ratio of number of points in E upon number of points in S), which satisfies the following axioms:

1. $P(E) \geq 0$
2. $P(S) = 1$
3. $P(E_1 \cup E_2) = P(E_1) + P(E_2)$ if $E_1 \cap E_2 = \varphi$

where E_1 and E_2 are two arbitrary events.

If the sample space is not finite and E_1, E_2 is an infinite sequence of mutually exclusive events in S, $E_i \cap E_j = \varphi$ for $i \neq j$, then

$$P\left(\bigcup_{i=1}^{\infty} E_i\right) = \sum_{i=1}^{\infty} P(E_i) \tag{3.1}$$

3.1.1 Elementary Properties of Probability

Using the same axioms, we have the following properties of probability
Let us consider two arbitrary events A and B, then

1. $P(A^c) = 1 - P(A)$ is the complement of event A. $\qquad$ (3.2)

2. $P(\phi) = 0$. $\qquad$ (3.3)

3. $P(A) \le P(B)$ if $A \subseteq B$. $\qquad$ (3.4)

4. $P(A) \le 1$. $\qquad$ (3.5)

5. $P(A \cup B) = P(A) + P(B) - P(A \cap B)$. $\qquad$ (3.6)

6. If $A_1, A_2, \ldots, A_n$ are n arbitrary events in S, then

$$P\left(\bigcup_{i=1}^{n} A_i\right) = \sum_{i=1}^{n} P(A_i) - \sum_{i \ne j} P(A_i \cap A_j) + \sum_{i \ne j \ne k} P(A_i \cap A_j \cap A_k)$$
$$- \cdots (-1)^{n-1} P(A_1 \cap A_2 \cap \cdots \cap A_n) \qquad (3.7)$$

7. For finite exclusive events $A_1, A_2, \ldots, A_n$ in S ($A_i \cap A_j = \phi$ for $i \ne j$), then

$$P\left(\bigcup_{i=1}^{n} A_i\right) = \sum_{i=1}^{n} P(A_i) \qquad (3.8)$$

3.1.2 Conditional Probability

The conditional probability of an event A given the event B is denoted by $P(A|B)$, which is defined as

$$P(A|B) = \frac{P(A \cap B)}{P(B)}, \quad P(B) > 0 \qquad (3.9)$$

where $P(A \cap B)$ is the joint probability of A and B.
Similarly, the conditional probability of an event A given the event B is

$$P(B|A) = \frac{P(A \cap B)}{P(A)}, \quad P(A) > 0 \qquad (3.10)$$

From Equations 3.9 and 3.10, we have

$$P(A \cap B) = P(A|B)P(B) = P(B|A)P(A) \qquad (3.11)$$

Equation 3.11 is often quite useful to compute the joint probability and we obtain the Bayes rule as follows:

$$P(A|B) = \frac{P(B|A)P(A)}{P(B)} \tag{3.12}$$

3.1.2.1 Random Variables

Consider a sample space S of a random experiment. Then a *random variable* $X(\tau)$ is a single real-valued function defined on the sample space S. For every number x, the probability is

$$P(X = x) \tag{3.13}$$

where $X(\tau)$ assumes x is defined.

Generally, we use a single letter X for the function $X(\tau)$. The sample space S is the domain of the random variable X and the collection of all the numbers (values of X) is the range of the random variable X. Here, it may be noted that two or more different sample points might give the same value of $X(\tau)$, but two or more different numbers in the range cannot be assigned to the same point.

If X is a random variable and x is a fixed real number, then we may define the event $(X = x)$ as

$$(X = x) = \{\tau : X(\tau) = x\} \tag{3.14}$$

For fixed real numbers x, x_1 and x_2, we can define the events in the following way:

$$(X < x) = \{\tau : X(\tau) < x\}$$
$$(X \geq x) = \{\tau : X(\tau) \geq x\} \tag{3.15}$$
$$(x_1 < X \leq x_2) = \{\tau : x_1 < X(\tau) \leq x_2\}$$

The probabilities for the events in Equations 3.14 and 3.15 are denoted as

$$P(X = x) = P\{\tau : X(\tau) = x\}$$
$$P(X < x) = P\{\tau : X(\tau) < x\}$$
$$P(X \geq x) = P\{\tau : X(\tau) \geq x\} \tag{3.16}$$
$$P(x_1 < X \leq x_2) = P\{\tau : x_1 < X(\tau) \leq x_2\}$$

A random variable and its distributions are of two types such as discrete and continuous. A random variable X and its distribution is called discrete if X considers only finitely many or countable values $x_1, x_2, x_3, \ldots$, of X, with probabilities. $p_1 = P(X = x_1)$, $p_2 = P(X = x_2)$, $p_3 = P(X = x_3)$, ...

The distribution of X is obtained by the probability function $f(x)$ of X as

$$f(x) = \begin{cases} p_i & \text{if } x = x_i, \ i = 1, \ 2, \ \ldots \\ 0 & \text{otherwise} \end{cases} \tag{3.17}$$

From (3.17), we get the values of the distribution function $F(x)$ as

$$F(x) = \sum_{x_i \leq x} f(x_i) = \sum_{x_i \leq x} p(x_i).$$
(3.18)

This is a staircase or step function at the possible values x_i of X are constant in between.
Properties of $p(x)$:

1. $0 \leq p(x_i) \leq 1, i = 1, 2, \ldots;$
2. $p(x) = 0$, if $x \neq x_i$ for $i = 1, 2, \ldots;$
3. $\sum_i p(x_i) = 1.$

Consider an experiment of tossing a fair coin thrice. The sample space S has a total of eight sample points, that is $S = \{HHH, HHT, HTH, THH, HTT, THT, TTH, TTT\}$, where H and T are heads and tails of the coin. If X is the random variable giving the heads obtained, then

$$(X = 1) = \{\tau : X(\tau) = 1\} = \{HTT, THT, TTH\};$$

$$(X = 2) = \{\tau : X(\tau) = 2\} = \{HHT, HTH, THH\};$$

$$(X \leq 2) = \{\tau : X(\tau) \leq 2\} = \{TTT, HTT, THT, TTH, HHT, HTH, THH\}.$$

The probabilities for these random variables are

$$P(X = 1) = P\{\tau : X(\tau) = 1\} = \frac{3}{8};$$

$$P(X = 2) = P\{\tau : X(\tau) = 2\} = \frac{3}{8};$$

$$P(X \leq 2) = P\{\tau : X(\tau) \leq 2\} = \frac{7}{8}.$$

A random variable X and its distribution are called continuous if in the distribution of X, F can be given by an integral of the following:

$$\int_{-\infty}^{\infty} f(\xi) d\xi.$$
(3.19)

If $F(x)$ is the distribution function, then

$$F(x) = \int_{-\infty}^{x} f(\xi) d\xi.$$

That means for every value of x and $f(x)$ is continuous, the differentiation of $F(x)$ gives $f(x)$.
For an interval $a < x \leq b$, $P(a < X \leq b) = F(b) - F(a) = \int_a^b f(\xi) d\xi.$

Properties of $f(x)$:

1. $f(x) > 0$;
2. $\int_{-\infty}^{\infty} f(x) dx = 1$;
3. $f(x)$ is piecewise continuous;
4. $P(a < X \le b) = \int_{a}^{b} f(x) dx$.

3.1.2.2 Mean and Variance of a Distribution

The mean or expectation of X, denoted by μ_X or $E(X)$, is defined as

$$\mu_X = E(X) = \sum_i x_i p(x_i), \quad \text{for discrete distribution};$$

$$\mu_X = E(X) = \int_{-\infty}^{\infty} x f(x) dx, \quad \text{for continuous distribution}.$$

(3.20)

The variance σ^2 is defined as

$$\sigma_X^2 = E\left[\{X - E(X)\}^2\right] = \sum_i (x_i - \mu_X)^2 p(x_i), \quad \text{for discrete distribution};$$

$$\sigma_X^2 = E\left[\{X - E(X)\}^2\right] = \int_{-\infty}^{\infty} (x - \mu_X)^2 f(x) dx, \quad \text{for continuous distribution}.$$

(3.21)

The nth moment of the random variable X is

$$E(X^n) = \sum_i x_i^n p(x_i), \quad \text{for discrete distribution};$$

$$E(X^n) = \int_{-\infty}^{\infty} x^n f(x) dx, \quad \text{for continuous distribution}.$$

(3.22)

3.1.3 Distributions

There are various distributions and these are of discrete or continuous type. Some of the standard distributions are as follows.

3.2 Binomial Distribution

This type of distribution occurs in games of chance, opinion polls, etc. If our aim is to know the times of the occurrence of an event A in n trials, then $P(A) = p$ is the success (probability of occurrence of an event A). The probability of not occurring an event A is $q = 1 - p$.

Here $X=x$ means that A occurs in x trials and in $n-x$ trials it does not occur. As such, the probability function will be

$$f(x) = \binom{n}{x} p^x q^{n-x}, \quad x = 0, 1, 2, \ldots, n. \tag{3.23}$$

where $0 \leq p \leq 1$ and $\binom{n}{x} = \dfrac{n!}{x!(n-x)!}$ which is called the binomial coefficient.

The distribution function $F(x)$ is defined as

$$F(x) = \sum_{x=0}^{n} \binom{n}{x} p^x q^{n-x}, \quad n \leq x < n+1 \tag{3.24}$$

The mean and variance of the binomial distribution are $\mu_X = np$ and $\sigma_X^2 = npq$.

3.3 Poisson Distribution

The Poisson distribution is a discrete distribution with the probability function

$$f(x) = \frac{\lambda^x}{x!} e^{-\lambda}, \quad x = 0, 1, 2, \ldots. \tag{3.25}$$

The corresponding distribution function is

$$F(x) = e^{-\lambda} \sum_{x=0}^{n} \frac{\lambda^x}{x!}, \quad n \leq x < n+1. \tag{3.26}$$

This distribution is a special case of binomial distribution, where $p \to 0$ and $n \to \infty$. The mean and variance of the binomial distribution are $\mu_X = \lambda$ and $\sigma_X^2 = \lambda$.

3.4 Normal or Gaussian Distribution

This is a continuous distribution and its probability function is defined as

$$f(x) = \frac{1}{\sigma\sqrt{2\pi}} e^{-(x-\mu)^2/2\sigma^2}. \tag{3.27}$$

The distribution function of the normal distribution is

$$F(x) = \frac{1}{\sigma\sqrt{2\pi}} \int_{-\infty}^{x} e^{-(\xi-\mu)^2/2\sigma^2} d\xi. \tag{3.28}$$

The integral (3.28) may be written in the closed form as

$$F(x) = \Phi\left(\frac{x-\mu}{\sigma}\right) \tag{3.29}$$

where $\Phi(z) = \dfrac{1}{\sqrt{2\pi}} \int e^{-\xi^2/2}\, d\xi$ and the mean and variance of the normal distribution is $\mu_X = \mu$ and $\sigma_X^2 = \sigma^2$, respectively.

3.4.1 Stochastic Differential Equations

It is well known that the straight-line segments are the backbone of the differential calculus. Moreover, differentiable functions are the basics of differential calculus. The behaviour may be locally approximated by straight-line segments. The same idea has been adapted in various methods, such as the Euler method for approximating differentiable functions defined by differential equations.

But in the case of the Brownian motion, the notion of a straight line produces another image of the Brownian motion. The mentioned self-similarity is ideal for an infinitesimal building block, and we can make the global Brownian motion out of lots of local Brownian motions. As such, one may build other stochastic processes out of the suitably scaled Brownian motion. Furthermore, if one may consider straight-line segments, then one can overlay the behaviour of differentiable functions onto the stochastic processes as well. Thus, straight-line segments and the local Brownian motion are the backbone of stochastic calculus.

With stochastic differential calculus, one can make new stochastic processes. We can make new stochastic processes by specifying the base deterministic function, the straight line, the base stochastic process and the standard Brownian motion. As such, the local change in the value of the stochastic process over a time interval of (infinitesimal) length dt as

$$dX(t) = a(t,X)dt + b(t,X)dW_t, \quad X(t_0) = X_0 \tag{3.30}$$

In the stochastic differential equation, the initial point (t_0, X_0) is specified, possibly with a random variable X_0 for a given distribution. A deterministic component at each point has a slope determined through a at that point. In addition, some random perturbation affects the evolution of the process. The random perturbation is normally distributed with mean zero. The variance of the random perturbation is $(b(t, X))^2$ at $(t, X(t))$. This is a simple expression of a stochastic differential equation (SDE), which determines a stochastic process, just as an ordinary differential equation (ODE) determines a differentiable function. We extend this process with the incremental change information and repeat. This is an expression in words of the Euler–Maruyama method for numerically simulating the stochastic differential expression.

3.4.1.1 *Euler–Maruyama and Milstein Methods*

Numerical methods for the SDEs are well known. Accordingly, one may consider the Euler–Maruyama method, which is used to solve the said uncertain problems. As such, let us assign a grid of points, $c = t_0 < t_1 < t_2 < \cdots < t_{n-1} < t_n = d$ and approximate x values $w_0 < w_1 < w_2 < \cdots < w_n$ to be determined at the respective t points.

Let us consider SDE initial value problem (Black and Scholes 1973)

$$\begin{cases} dX(t) = a(t,X)dt + b(t,X)dW_t \\ X(0) = X_0 \end{cases} \tag{3.31}$$

As said earlier, numerical schemes for Equation 3.31 have been incorporated for the two well-known methods (i.e. Euler–Maruyama and Milstein), which is discussed next.

3.4.1.1.1 Euler–Maruyama Method

We take a time-discrete approximation of the SDE (Higham 2001)

$$\begin{cases} dX(t) = a(t,X)dt + b(t,X)dW_t \\ X(c) = X_c \end{cases} \tag{3.32}$$

Then the approximation scheme for Euler–Maruyama may be represented as follows (Sauer 2012):

$$\begin{aligned} X_0 &= w_0 \\ w_{i+1} &= w_i + a(t_i,w_i)\Delta t_{i+1} + b(t_i,w_i)\Delta W_i \end{aligned} \tag{3.33}$$

where X_c is the value of X at $t = c$,

$$\Delta t_{i+1} = t_{i+1} - t_i,$$
$$\Delta W_{i+1} = W(t_{i+1}) - W(t_i).$$

We define $N(0, 1)$ to be the normal distribution and each random number ΔW_i is computed as

$$\Delta W_i = z_i \sqrt{\Delta t_i}, \quad \text{where } z_i \in N(0, 1).$$

The obtained set $\{w_0, w_1, ..., w_n\}$ is an approximation realization of the stochastic solution $X(t)$, which depends on the random number z_i that was chosen. Since W_t is a stochastic process, each realization will be different and so will our approximations.

3.4.1.1.2 Milstein Method

The approximation scheme of the Milstein method for Equation 3.32 may be written in the following way (Sauer 2012):

$$\begin{aligned} X_0 &= w_0 \\ w_{i+1} &= w_i + a(t_i,w_i)\Delta t_{i+1} + b(t_i,w_i)\Delta W_i + \frac{1}{2}b(t_i,w_i)\frac{\partial b}{\partial x}(t_i,w_i)\left(\Delta W_i^2 - \Delta t_i\right) \end{aligned} \tag{3.34}$$

3.5 Interval Uncertainty

The closed interval $[a, b]$ is the set of real numbers that is represented as

$$[a, b] = \{x \in R : a \le x \le b\}. \tag{3.35}$$

Here, small letters are used to denote the intervals and their endpoints. The left and right endpoints of an interval x will be denoted by $\underline{x}$ and $\bar{x}$, respectively. Thus, $x = \left[\underline{x}, \bar{x}\right]$.

The intervals x and y are said to be equal if they are the same sets. Operationally, this happens if their corresponding endpoints are equal:

$$x = y \quad \text{if } \underline{x} = \underline{y} \text{ and } \overline{x} = \overline{y}.$$

3.5.1 Degenerate Intervals

We say that x is degenerate if $\underline{x} = \overline{x}$. Such an interval contains a single real number x. In this sense, we may write equations as $0 = [0, 0]$.

3.5.2 Intersection, Union and Interval Hull

The intersection of two intervals x and y is empty if either $\overline{y} < \underline{x}$ or $\overline{x} < \underline{y}$. We denote the empty set by ϕ and write $x \cap y = \phi$, indicating that x and y have no points in common. Otherwise, we may define the intersection $x \cap y$ as the interval

$$x \cap y = \{z : z \in x \text{ and } z \in y\}$$
$$= \left[\max\{\underline{x}, \underline{y}\}, \ \min\{\overline{x}, \overline{y}\} \right] \tag{3.36}$$

In the latter case, the union of x and y is also an interval

$$x \cup y = \{z : z \in x \text{ or } z \in y\}$$
$$= \left[\min\{\underline{x}, \underline{y}\}, \ \max\{\overline{x}, \overline{y}\} \right] \tag{3.37}$$

In general, the union of two intervals may not be an interval. However, the interval hull of two intervals, defined by

$$x \ \underline{\cup} \ y = \left[\min\{\underline{x}, \underline{y}\}, \ \max\{\overline{x}, \overline{y}\} \right], \tag{3.38}$$

is always an interval and can be used in interval computations. We have $x \cup y \subseteq x \ \underline{\cup} \ y$ for any two intervals x and y.

3.5.3 Width, Absolute Value and Midpoint

The width of an interval x is defined and denoted by $w(x) = \overline{x} - \underline{x}$.

The absolute value of x, denoted $|x|$, is the maximum of the absolute values of its endpoints

$$|x| = \max\{|\underline{x}|, |\overline{x}|\}. \tag{3.39}$$

The midpoint of x is given by $m(x) = \dfrac{1}{2}(\underline{x} + \overline{x})$.

3.5.4 Interval Arithmetic

If $[\underline{x}, \overline{x}]$ and $[\underline{y}, \overline{y}]$ are two intervals, then the interval arithmetic (Neumaier 1990; Moore et al. 2014) may be written as

$$[\underline{x}, \overline{x}] + [\underline{y}, \overline{y}] = [\underline{x} + \underline{y}, \ \overline{x} + \overline{y}] \tag{3.40}$$

$$[\underline{x}, \overline{x}] - [\underline{y}, \overline{y}] = [\underline{x} - \overline{y}, \ \overline{x} - \underline{y}] \tag{3.41}$$

$$[\underline{x}, \overline{x}] \times [\underline{y}, \overline{y}] = \left[\min\{\underline{xy}, \underline{x}\overline{y}, \overline{x}\underline{y}, \overline{xy}\}, \ \max\{\underline{xy}, \underline{x}\overline{y}, \overline{x}\underline{y}, \overline{xy}\} \right] \tag{3.42}$$

$$[\underline{x}, \overline{x}] \div [\underline{y}, \overline{y}] = \left[\min\{\underline{x} \div \underline{y}, \ \underline{x} \div \overline{y}, \ \overline{x} \div \underline{y}, \ \overline{x} \div \overline{y}\}, \right.$$
$$\left. \max\{\underline{x} \div \underline{y}, \ \underline{x} \div \overline{y}, \ \overline{x} \div \underline{y}, \ \overline{x} \div \overline{y}\} \right], \quad 0 \notin [\underline{y}, \overline{y}] \tag{3.43}$$

The traditional interval arithmetic is sometimes difficult to use. When large numbers of computations are involved, then the process becomes difficult to handle and uncertainty rises. It may also be difficult to formulate the methods in general. Here, the traditional interval arithmetics have been redefined and proposed.

Let us consider two intervals $[\underline{x}, \overline{x}]$ and $[\underline{y}, \overline{y}]$, then the traditional interval arithmetic may be represented in an alternate form as follows.

If all the values of the interval are in R^+ or R^-, then the arithmetic rules may be written as

$$[\underline{x}, \overline{x}] + [\underline{y}, \overline{y}] = \left[\min\left\{\lim_{n \to \infty} l_1 + \lim_{n \to \infty} l_2, \lim_{n \to 1} l_1 + \lim_{n \to 1} l_2\right\}, \max\left\{\lim_{n \to \infty} l_1 + \lim_{n \to \infty} l_2, \lim_{n \to 1} l_1 + \lim_{n \to 1} l_2\right\} \right] \tag{3.44}$$

$$[\underline{x}, \overline{x}] - [\underline{y}, \overline{y}] = \left[\min\left\{\lim_{n \to \infty} l_1 - \lim_{n \to 1} l_2, \lim_{n \to 1} l_1 - \lim_{n \to \infty} l_2\right\}, \max\left\{\lim_{n \to \infty} l_1 - \lim_{n \to 1} l_2, \lim_{n \to 1} l_1 - \lim_{n \to \infty} l_2\right\} \right] \tag{3.45}$$

$$[\underline{x}, \overline{x}] \times [\underline{y}, \overline{y}] = \left[\min\left\{\lim_{n \to \infty} l_1 \times \lim_{n \to \infty} l_2, \lim_{n \to 1} l_1 \times \lim_{n \to 1} l_2\right\}, \max\left\{\lim_{n \to \infty} l_1 \times \lim_{n \to \infty} l_2, \lim_{n \to 1} l_1 \times \lim_{n \to 1} l_2\right\} \right] \tag{3.46}$$

$$[\underline{x}, \overline{x}] \div [\underline{y}, \overline{y}] = \left[\min\left\{\lim_{n \to \infty} l_1 \div \lim_{n \to 1} l_2, \lim_{n \to 1} l_1 \div \lim_{n \to \infty} l_2\right\}, \max\left\{\lim_{n \to \infty} l_1 \div \lim_{n \to 1} l_2, \lim_{n \to 1} l_1 \div \lim_{n \to \infty} l_2\right\} \right] \tag{3.47}$$

where

$$l_1 = \underline{x} + \frac{\overline{x} - \underline{x}}{n},$$

$$\lim_{n \to \infty} l_1 = \lim_{n \to \infty} \underline{x} + \frac{\overline{x} - \underline{x}}{n} = \underline{x}$$

$$\lim_{n \to 1} l_1 = \lim_{n \to 1} \underline{x} + \frac{\overline{x} - \underline{x}}{n} = \overline{x}$$

$$l_2 = \underline{y} + \frac{\overline{y} - \underline{y}}{n},$$

$$\lim_{n \to \infty} l_2 = \lim_{n \to \infty} \underline{y} + \frac{\overline{y} - \underline{y}}{n} = \underline{y}, \quad \lim_{n \to 1} l_2 = \lim_{n \to 1} \underline{y} + \frac{\overline{y} - \underline{y}}{n} = \overline{y}.$$

One may observe that if we consider an interval that includes 0, then we will have at least one solution that is undefined in the division operation. For example, if we consider two intervals $[-1, 2]$ and $[-3, 1]$ and divide $[-1, 2]$ by $[-3, 1]$, then we have a solution where 0 divides 0, which is not defined in general. Hence, 0 has not been considered in the proposed arithmetic. However, this difficulty is a great challenge to the interval/fuzzy researchers.

3.6 Fuzzy Uncertainty

A classical or crisp set A can be defined as a collection of objects or elements of a universal set X. The elements of the set (say A) may be defined by using the characteristic function χ_A, which is

$\chi_A : X \rightarrow \{0, 1\}$, where X is the universal set.

χ_A indicates the membership of the element $x \in X$ if $\chi_A(x) = 1$ and its non-membership if $\chi_A(x) = 0$.

3.6.1 Definitions

In the following, some important definitions related to fuzzy sets are introduced and explained.

3.6.1.1 Fuzzy Set

If X is a collection of objects or elements (denoted by x), then a fuzzy set $\widetilde{A}$ in X is a set of ordered pairs:

$$\widetilde{A} = \left\{ \left(x, \mu_{\widetilde{A}}(x) \right) \mid x \in X \right\} \tag{3.48}$$

where $\mu_{\widetilde{A}}(x)$ is the membership function of x.

> **Example 3.1**
>
> Let us consider a fuzzy set $\widetilde{A}$ = real numbers larger than 15 (Zimmermann 1991) as $\widetilde{A} = \left\{ \left(x, \mu_{\widetilde{A}}(x) \right) \mid x \in X \right\}$, where
>
> $$\mu_{\widetilde{A}}(x) = \begin{cases} 0, & x \le 15 \\ \left(1 + \dfrac{1}{(x-15)^2} \right)^{-1}, & x > 15 \end{cases} \tag{3.49}$$

3.6.1.2 Support of a Fuzzy Set

The support of a fuzzy set $\widetilde{A}$ is the crisp set of elements $x \in X$ that has non-zero membership grades in $\widetilde{A}$.

The support of a fuzzy set $\tilde{A}$ may be written as

$$\text{supp}(\tilde{A}) = \{x \in X | \mu_{\tilde{A}}(x) > 0\} \tag{3.50}$$

3.6.1.3 α-*Level Set of a Fuzzy Set*

The α-level set A_α of a fuzzy set $\tilde{A}$ is the crisp set of all elements $x \in X$ that belongs to the fuzzy set $\tilde{A}$ at least to the degree $\alpha \in [0, \ 1]$.

$$A_\alpha = \{x \in X | \mu_{\tilde{A}}(x) \geq \alpha\} \tag{3.51}$$

The α-level set $A_{\alpha+}$ with

$$A_{\alpha+} = \{x \in X | \mu_{\tilde{A}}(x) > \alpha\} \tag{3.52}$$

is called strong α-level set of the fuzzy set $\tilde{A}$.

3.6.1.4 Convexity of a Fuzzy Set

A fuzzy set $\tilde{A}$ is convex if

$$\mu_{\tilde{A}}(\lambda x_1 + (1-\lambda)x_2) \geq \min(\mu_{\tilde{A}}(x_1), \ \mu_{\tilde{A}}(x_2)), \quad x_1, x_2 \in X, \ \lambda \in [0, 1] \tag{3.53}$$

In other words, a fuzzy set is convex if all α-level sets are convex.

3.6.1.5 Height of a Fuzzy Set

The height $h(\tilde{A})$ of a fuzzy set $\tilde{A}$ is the largest membership grade obtained by any element in that set.

$$h(\tilde{A}) = \sup_{x \in X} \tilde{A}(x) \tag{3.54}$$

A fuzzy set $\tilde{A}$ is called normal when $h(\tilde{A}) = 1$ and subnormal if $h(\tilde{A}) < 1$.

3.6.2 Fuzzy Numbers

A fuzzy set $\tilde{A}$ is called a fuzzy number if it satisfies the following conditions:

1. $\tilde{A}$ is normal, that is $h(\tilde{A}) = 1$.
2. $\tilde{A}$ is convex.
3. The membership function $\mu_{\tilde{A}}(x)$ is at least piecewise continuous.

3.6.2.1 Triangular Fuzzy Number

A fuzzy number $\tilde{A} = \left[a^L, a^N, a^R \right]$ (Figure 3.1) is said to be a triangular fuzzy number (TFN) when the membership function is given by

$$\mu_{\tilde{A}}(x) = \begin{cases} 0, & x \leq a^L; \\ \dfrac{x - a^L}{a^N - a^L}, & a^L \leq x \leq a^N; \\ \dfrac{a^R - x}{a^R - a^N}, & a^N \leq x \leq a^R; \\ 0, & x \geq a^R. \end{cases}$$

The TFN $\tilde{A} = \left[a^L, a^N, a^R \right]$ may be expressed into an ordered pair function by using α-cut as follows:

$$\left[\underline{f}(\alpha), \overline{f}(\alpha) \right] = \left[a^L + \left(a^N - a^L \right)\alpha, \ a^R - \left(a^R - a^N \right)\alpha \right], \quad \alpha \in [0,\ 1]$$

3.6.2.2 Trapezoidal Fuzzy Number

A fuzzy number $\tilde{A} = \left[a^L, a^{NL}, a^{NR}, a^R \right]$ (Figure 3.2) is said to be trapezoidal fuzzy number (TRFN) when the membership function is given by

$$\mu_{\tilde{A}}(x) = \begin{cases} 0, & x \leq a^L; \\ \dfrac{x - a^L}{a^{NL} - a^L}, & a^L \leq x \leq a^{NL}; \\ 1, & a^{NL} \leq x \leq a^{NR}; \\ \dfrac{a^R - x}{a^R - a^{NR}}, & a^{NR} \leq x \leq a^R; \\ 0, & x \geq a^R. \end{cases}$$

Again, the TRFN may be expressed into an ordered pair function through α-cut in the following manner:

$$\left[\underline{g}(\alpha), \overline{g}(\alpha) \right] = \left[a^L + \left(a^{NL} - a^L \right), \ a^R - \left(a^R - a^{NR} \right) \right], \quad \alpha \in [0,\ 1]$$

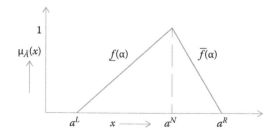

FIGURE 3.1
Triangular fuzzy number (TFN).

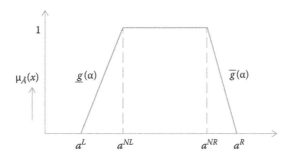

FIGURE 3.2
Trapezoidal fuzzy number (TRFN).

3.6.2.3 Fuzzy Arithmetic

Let us consider $\left[\underline{x}(\alpha),\ \bar{x}(\alpha)\right]$ and $\left[\underline{y}(\alpha),\ \bar{y}(\alpha)\right]$ to be two fuzzy numbers, then the fuzzy arithmetic (Zimmermann 1991; Hanss 2005) may be written as

$$\left[\underline{x}(\alpha),\ \bar{x}(\alpha)\right]+\left[\underline{y}(\alpha),\ \bar{y}(\alpha)\right]=\left[\underline{x}(\alpha)+\underline{y}(\alpha),\ \bar{x}(\alpha)+\bar{y}(\alpha)\right] \tag{3.55}$$

$$\left[\underline{x}(\alpha),\ \bar{x}(\alpha)\right]-\left[\underline{y}(\alpha),\ \bar{y}(\alpha)\right]=\left[\underline{x}(\alpha)-\bar{y}(\alpha),\ \bar{x}(\alpha)-\underline{y}(\alpha)\right] \tag{3.56}$$

$$\left[\underline{x}(\alpha),\ \bar{x}(\alpha)\right]\times\left[\underline{y}(\alpha),\ \bar{y}(\alpha)\right]=\left[\min\{\underline{x}(\alpha)\underline{y}(\alpha),\ \underline{x}(\alpha)\bar{y}(\alpha),\ \bar{x}(\alpha)\underline{y}(\alpha),\ \bar{x}(\alpha)\bar{y}(\alpha)\},\right.$$
$$\left.\max\{\underline{x}(\alpha)\underline{y}(\alpha),\ \underline{x}(\alpha)\bar{y}(\alpha),\ \bar{x}(\alpha)\underline{y}(\alpha),\ \bar{x}(\alpha)\bar{y}(\alpha)\}\right] \tag{3.57}$$

$$\left[\underline{x}(\alpha),\ \bar{x}(\alpha)\right]\div\left[\underline{y}(\alpha),\ \bar{y}(\alpha)\right]=\left[\min\{\underline{x}(\alpha)/\underline{y}(\alpha),\ \underline{x}(\alpha)/\bar{y}(\alpha),\ \bar{x}(\alpha)/\underline{y}(\alpha),\ \bar{x}(\alpha)/\bar{y}(\alpha)\},\right.$$
$$\left.\max\{\underline{x}(\alpha)/\underline{y}(\alpha),\ \underline{x}(\alpha)/\bar{y}(\alpha),\ \bar{x}(\alpha)/\underline{y}(\alpha),\ \bar{x}(\alpha)/\bar{y}(\alpha)\}\right] \tag{3.58}$$

From this discussion, we may say that the intervals are the special cases of fuzzy numbers. For each membership value (α), we get an interval. In this regard, if we consider a TFN $\tilde{A}=\left[a^{L},a^{N},a^{R}\right]$, then the same may be represented by using α-cut as follows:

$$\left[\underline{f}(\alpha),\bar{f}(\alpha)\right]=\left[a^{L}+\left(a^{N}-a^{L}\right)\alpha,\ a^{R}-\left(a^{R}-a^{N}\right)\alpha\right],\quad \alpha\in\left[0,\ 1\right].$$

If we take $\alpha=0$, then we have the interval $[a^{L},\ a^{R}]$ and at $\alpha=1$, the interval $[a^{N},\ a^{N}]$ becomes crisp.

As such, a TFN $\tilde{A}=\left[a^{L},a^{N},a^{R}\right]$ may be represented into an ordered pair (interval) form by using α-cut as follows:

$$\tilde{A}=\left[a^{L},\ a^{N},\ a^{R}\right]=\left[a^{L}+\left(a^{N}-a^{L}\right)\alpha,\ a^{R}-\left(a^{R}-a^{N}\right)\alpha\right]$$
$$=\left[\underline{f}(\alpha),\ \bar{f}(\alpha)\right],\quad \alpha\in\left[0,\ 1\right]$$

Here, $\underline{f}(\alpha)$ and $\overline{f}(\alpha)$ are monotonic increasing and decreasing functions, respectively. Using these functions, we have modified the interval arithmetic (α-cut of fuzzy numbers) as follows:

$$\left[\underline{x}(\alpha),\ \overline{x}(\alpha)\right]+\left[\underline{y}(\alpha),\ \overline{y}(\alpha)\right]=\left[\min\left\{\lim_{n\to\infty} m_1 + \lim_{n\to\infty} m_2,\ \lim_{n\to 1} m_1 + \lim_{n\to 1} m_2\right\},\right.$$
$$\left.\max\left\{\lim_{n\to\infty} m_1 + \lim_{n\to\infty} m_2,\ \lim_{n\to 1} m_1 + \lim_{n\to 1} m_2\right\}\right] \tag{3.59}$$

$$\left[\underline{x}(\alpha),\ \overline{x}(\alpha)\right]-\left[\underline{y}(\alpha),\ \overline{y}(\alpha)\right]=\left[\min\left\{\lim_{n\to\infty} m_1 - \lim_{n\to 1} m_2,\ \lim_{n\to 1} m_1 - \lim_{n\to\infty} m_2\right\},\right.$$
$$\left.\max\left\{\lim_{n\to\infty} m_1 - \lim_{n\to 1} m_2,\ \lim_{n\to 1} m_1 - \lim_{n\to\infty} m_2\right\}\right] \tag{3.60}$$

$$\left[\underline{x}(\alpha),\ \overline{x}(\alpha)\right]\times\left[\underline{y}(\alpha),\ \overline{y}(\alpha)\right]=\left[\min\left\{\lim_{n\to\infty} m_1 \times \lim_{n\to\infty} m_2,\ \lim_{n\to 1} m_1 \times \lim_{n\to 1} m_2\right\},\right.$$
$$\left.\max\left\{\lim_{n\to\infty} m_1 \times \lim_{n\to\infty} m_2,\ \lim_{n\to 1} m_1 \times \lim_{n\to 1} m_2\right\}\right] \tag{3.61}$$

$$\left[\underline{x}(\alpha),\ \overline{x}(\alpha)\right]\div\left[\underline{y}(\alpha),\ \overline{y}(\alpha)\right]=\left[\min\left\{\lim_{n\to\infty} m_1 \div \lim_{n\to 1} m_2,\ \lim_{n\to 1} m_1 \div \lim_{n\to\infty} m_2\right\},\right.$$
$$\left.\max\left\{\lim_{n\to\infty} m_1 \div \lim_{n\to 1} m_2,\ \lim_{n\to 1} m_1 \div \lim_{n\to\infty} m_2\right\}\right] \tag{3.62}$$

where

$$m_1 = \underline{x}(\alpha) + \frac{\overline{x}(\alpha) - \underline{x}(\alpha)}{n}$$

$$m_2 = \underline{y}(\alpha) + \frac{\overline{y}(\alpha) - \underline{y}(\alpha)}{n},\quad n \in [1,\infty).$$

3.6.3 Interval and Fuzzy Systems of Equations

The linear system

$$a_{11}x_1 + a_{12}x_2 + \cdots + a_{1n}x_n = y_1$$
$$a_{21}x_1 + a_{22}x_2 + \cdots + a_{2n}x_n = y_2$$
$$\vdots \tag{3.63}$$
$$a_{n1}x_1 + a_{n2}x_2 + \cdots + a_{nn}x_n = y_n$$

where
  the coefficient matrix $A = (a_{ij})$, $1 \le i \le n$; $1 \le j \le n$ is an $n \times n$ crisp matrix
  y_i, $1 \le i \le n$ are interval numbers, is called an interval linear system

If the coefficient matrix $A = (a_{ij})$ and y_i, $1 \le i \le n$ both are interval then the system is called a fully interval linear system, whereas in Equation 3.63, if the coefficient matrix $A = (a_{ij})$, $1 \le i \le n$; $1 \le j \le n$ is an $n \times n$ crisp matrix and y_i, $1 \le i \le n$ are fuzzy numbers, it is called a fuzzy linear system (FLS). If the coefficient matrix $A = (a_{ij})$ and y_i, $1 \le i \le n$ both are fuzzy then the system is called fully fuzzy linear system (FFLS).

3.6.3.1 Linear System of Equations with Triangular Fuzzy Numbers

If the coefficient matrix $A = (a_{ij})$, $1 \leq i \leq n$ and $1 \leq j \leq n$ is an $n \times n$ crisp matrix and y_i, $1 \leq i \leq n$ are TFN, in Equation 3.63 we have an FLS with TFN, whereas if the coefficient matrix $A = (a_{ij})$ and y_i, $1 \leq i \leq n$ both are TFN, then the system is called FFLS with TFN. Now, an algorithm is proposed to solve linear system of equations with TFN.

Algorithm 3.1

Step 1. TFN is written in α-cut form.
   Let $[a^L, a^N, a^R]$ be a triangular fuzzy number, then it may be represented as

$$\left[a^L,\ a^N,\ a^R\right] = \left[a^L + \left(a^N - a^L\right)\alpha,\ a^R - \left(a^R - a^N\right)\alpha\right] = \left[\underline{f}(\alpha), \overline{f}(\alpha)\right].$$

Step 2. Now, the intervals are transferred into the crisp form using the transformation given in Equations 3.44 through 3.47.
Step 3. We get a system of linear equations with crisp values. This system may be solved by any standard method used for crisp values.
Step 4. Finally, the solution vector is

$$x = \left(\left[\lim_{n \to \infty} x_1(\alpha), \lim_{n \to 1} x_1(\alpha)\right], \left[\lim_{n \to \infty} x_2(\alpha),\ \lim_{n \to 1} x_2(\alpha)\right], \cdots, \left[\lim_{n \to \infty} x_n(\alpha), \lim_{n \to 1} x_n(\alpha)\right]\right)^T.$$

Example 3.2

Let us consider the following TFN system of linear equations:

$$\begin{aligned}
2x_1 + x_2 + 3x_3 &= \left[11,\ 19,\ 27\right] \\
4x_1 + x_2 - x_3 &= \left[-23,\ -13,\ -5\right] \\
-x_1 + 3x_2 + x_3 &= \left[10,\ 15,\ 27\right]
\end{aligned} \tag{3.64}$$

Equation 3.64 may be transformed into the following interval form:

$$\begin{aligned}
2x_1 + x_2 + 3x_3 &= \left[11 + 8\alpha,\ 27 - 8\alpha\right] \\
4x_1 + x_2 - x_3 &= \left[-23 + 10\alpha,\ -5 - 8\alpha\right] \\
-x_1 + 3x_2 + x_3 &= \left[10 + 5\alpha,\ 27 - 12\alpha\right]
\end{aligned} \tag{3.65}$$

The preceding equations may be written as

$$\begin{aligned}
2x_1 + x_2 + 3x_3 &= \left[\lim_{n \to \infty}\left\{(11 + 8\alpha) + \frac{16 - 16r}{n}\right\},\ \lim_{n \to 1}\left\{(11 + 8\alpha) + \frac{16 - 16r}{n}\right\}\right] \\
4x_1 + x_2 - x_3 &= \left[\lim_{n \to \infty}\left\{(-23 + 10\alpha) + \frac{18 - 18r}{n}\right\},\ \lim_{n \to 1}\left\{(-23 + 10\alpha) + \frac{18 - 18r}{n}\right\}\right] \\
-x_1 + 3x_2 + x_3 &= \left[\lim_{n \to \infty}\left\{(10 + 5\alpha) + \frac{17 - 17r}{n}\right\},\ \lim_{n \to 1}\left\{(10 + 5\alpha) + \frac{17 - 17r}{n}\right\}\right]
\end{aligned} \tag{3.66}$$

TABLE 3.1

Comparison between the Fuzzy Solutions in Terms of α-Cuts.

	Algorithm 3.1		Matinfar et al. (2008)	
x	Left	Right	Left	Right
x_1	$-\dfrac{45}{11}+\dfrac{23}{11}\alpha$	$-\dfrac{10}{11}-\dfrac{12}{11}\alpha$	$-4+2\alpha$	$-1-\alpha$
x_2	$-\dfrac{2}{11}+\dfrac{4}{11}\alpha$	$\dfrac{68}{11}-\dfrac{46}{11}\alpha$	$1+\alpha$	$5-3\alpha$
x_3	$\dfrac{71}{11}+\dfrac{6}{11}\alpha$	$\dfrac{83}{11}-\dfrac{6}{11}\alpha$	$6+\alpha$	$8-\alpha$

Now, solving Equation 3.66, we may get the solution vector as

$$x=\left(\left[-\frac{45}{11}+\frac{23}{11}\alpha,-\frac{10}{11}-\frac{12}{11}\alpha\right],\left[-\frac{2}{11}+\frac{4}{11}\alpha,\frac{68}{11}-\frac{46}{11}\alpha\right],\left[\frac{71}{11}+\frac{6}{11}\alpha,\frac{83}{11}-\frac{6}{11}\alpha\right]\right)^{T}.$$

The obtained results are compared with the results (Matinfar et al. 2008) for different membership functions. The results are presented in Table 3.1.

Example 3.3

Next, let us consider the fully fuzzy system of linear equations $Ax=b$, where

$$A=\begin{bmatrix}[5,6,10] & [3,5,7] & [1,3,4]\\ [4,12,32] & [2,14,29] & [0,8,18]\\ [14,24,58] & [2,32,62] & [1,20,44]\end{bmatrix} \quad \text{and} \quad b=\begin{bmatrix}[1,2,3]\\ [0,1,2]\\ [3,5,7]\end{bmatrix}.$$

The matrix A and vector b may be transformed into the interval form using Algorithm 3.1 and accordingly, we get

$$A=\begin{bmatrix}[5+\alpha,\,10-4\alpha] & [3+2\alpha,\,7-2\alpha] & [1+2\alpha,\,4-\alpha]\\ [4+8\alpha,\,32-20\alpha] & [2+12\alpha,\,29-15\alpha] & [8\alpha,\,18-10\alpha]\\ [14+10\alpha,\,58-34\alpha] & [2+30\alpha,\,62-30\alpha] & [1+19\alpha,\,44-24\alpha]\end{bmatrix} \quad \text{and} \quad b=\begin{bmatrix}[1+\alpha,\,3-\alpha]\\ [\alpha,\,2-\alpha]\\ [3+2\alpha,\,7-2\alpha]\end{bmatrix}.$$

Finally, the solution vector is obtained as

$$x=\left([0.1818,\,0.7083,\,1.6538],\,[-4,\,-2.25,\,-0.3636],\,[1.1818,\,3,\,3.6154]\right)^{T}.$$

3.6.3.2 Linear System of Equations with Trapezoidal Fuzzy Numbers

Equation 3.63 with TRFN may be converted into left monotonically increasing and right monotonically decreasing continuous functions over [0, 1]. Then, it may be solved by the following algorithm.

Algorithm 3.2

Step 1. Convert the TRFN into the following form:

$$\left[a^{L},\,a^{NL},\,a^{NR},\,a^{R}\right]=\left[\underline{g}(\alpha),\,\overline{g}(\alpha)\right].$$

Step 2. Now, the intervals are transferred into the crisp form using the transformation given in Equations 3.44 through 3.47.

Step 3. We get a system of linear equations obtained with crisp values. This system may be solved by any standard method used for crisp values.

Step 4. Finally, the solution vector is

$$x = \left(\left[\lim_{n \to \infty} x_1(\alpha), \lim_{n \to 1} x_1(\alpha) \right], \left[\lim_{n \to \infty} x_2(\alpha), \lim_{n \to 1} x_2(\alpha) \right], \dots, \left[\lim_{n \to \infty} x_n(\alpha), \lim_{n \to 1} x_n(\alpha) \right] \right)^T.$$

Example 3.4

Here, we consider the following system of linear equations with the TRFN

$$\begin{aligned} x_1 - x_2 &= [-31, -1, 3, 30] \\ x_1 + 5x_2 &= [-65, 1, 13, 100] \end{aligned} \tag{3.67}$$

Equation 3.67 may be transferred into the following α-cut form:

$$\begin{aligned} x_1 - x_2 &= [-31 + 30\alpha, 30 - 27\alpha] \\ x_1 + 5x_2 &= [-65 + 66\alpha, 100 - 87\alpha] \end{aligned} \tag{3.68}$$

Using Algorithm 3.2, we get the following solution:

$$x = \left([-36.6667, -0.6667, 4.6667, 41.6667], [-5.6667, 0.333, 1.6667, 11.6667] \right)^T.$$

Example 3.5

Finally, the fully fuzzy system of linear equations with TRFN is considered:

$$\begin{aligned} \left[1, 3, 6, 8 \right] x_1 + \left[3, 4, 6, 8 \right] x_2 &= \left[7, 27, 66, 136 \right] \\ \left[-5, 1, 2, 4 \right] x_1 + \left[2, 4, 5, 7 \right] x_2 &= \left[-41, 17, 37, 92 \right] \end{aligned} \tag{3.69}$$

Again, with the help of the α-cut, Equation 3.69 becomes

$$\begin{aligned} \left[1 + 2\alpha, 8 - 2\alpha \right] x_1 + \left[3 + \alpha, 8 - 2\alpha \right] x_2 &= \left[7 + 20\alpha, 136 - 70\alpha \right] \\ \left[-5 + 6\alpha, 4 - 2\alpha \right] x_1 + \left[2 + 2\alpha, 7 - 2\alpha \right] x_2 &= \left[-41 + 66\alpha, 92 - 55\alpha \right] \end{aligned} \tag{3.70}$$

Finally, solving Equation 3.70 using Algorithm 3.2, we get the solution vector:

$$x = \left([8.0588, 5, 6, 9], [-0.3529, 3, 5, 8] \right)^T.$$

3.6.3.3 Fuzzy Eigenvalue Problems

If $\widetilde{A}$ is a square fuzzy matrix of order n, then the values of λ for which the equation

$$\widetilde{A}x = \widetilde{\lambda}x \tag{3.71}$$

has non-trivial solutions are called the fuzzy eigenvalues of $\widetilde{A}$.

3.6.3.3.1. Algorithm to Find Eigenvalues and Eigenvectors of a Matrix

Let us consider $\widetilde{A}$ is a square fuzzy (TFN) matrix of order n, which is represented as

$$\widetilde{A} = \begin{bmatrix} \left[a_{11}^L,\ a_{11}^N,\ a_{11}^R\right] & \left[a_{12}^L,\ a_{12}^N,\ a_{12}^R\right] & \cdots & \left[a_{1n}^L,\ a_{1n}^N,\ a_{1n}^R\right] \\ \left[a_{21}^L,\ a_{21}^N,\ a_{21}^R\right] & \left[a_{22}^L,\ a_{22}^N,\ a_{22}^R\right] & \cdots & \left[a_{2n}^L,\ a_{2n}^N,\ a_{2n}^R\right] \\ \vdots & \vdots & \cdots & \vdots \\ \left[a_{n1}^L,\ a_{n1}^N,\ a_{n1}^R\right] & \left[a_{n2}^L,\ a_{n2}^N,\ a_{n2}^R\right] & \cdots & \left[a_{nn}^L,\ a_{nn}^N,\ a_{nn}^R\right] \end{bmatrix} \tag{3.72}$$

Step 1. Convert the square matrix $\widetilde{A}$ in the following form as discussed in Equation 3.71

$$\widetilde{A} = \left[a_{ij}^\alpha\right] \tag{3.73}$$

for the membership function α.

Step 2. Now, the obtained crisp matrix is solved in a traditional method to get eigenvalues and eigenvectors. The eigenvalues and eigenvectors involve the membership functions α and t.

Step 3. For each α, we get a set of eigenvalues and eigenvectors depending on t. Using the limit method (Chakraverty and Nayak 2013), we can get minimum and maximum eigenvalues and eigenvectors for each α.

Step 4. Finally, the fuzzy eigenvalues and eigenvectors may be obtained in the following form:

$$\left[\min\left\{\lim_{t\to1}\widetilde{\lambda}_i, \lim_{t\to\infty}\widetilde{\lambda}_i\right\},\ \max\left\{\lim_{t\to1}\widetilde{\lambda}_i, \lim_{t\to\infty}\widetilde{\lambda}_i\right\}\right],\quad i = 1,\ 2,\ \ldots,\ n$$

$$\left[\min\left\{\lim_{t\to1}\widetilde{v}_i, \lim_{t\to\infty}\widetilde{v}_i\right\},\ \max\left\{\lim_{t\to1}\widetilde{v}_i, \lim_{t\to\infty}\widetilde{v}_i\right\}\right],\quad i = 1,\ 2,\ \ldots,\ n.$$

Example 3.6

Let us consider the standard fuzzy eigenvalue problem $\widetilde{A}\widetilde{X} = \widetilde{\lambda}\widetilde{X}$.

$$\widetilde{A} = \begin{bmatrix} [1,\ 2,\ 3] & [0,\ 1,\ 3] \\ [7,\ 12,\ 14] & [1,\ 3,\ 8] \end{bmatrix}.$$

Now, applying this algorithm, the matrix $\widetilde{A}$ becomes

$$\widetilde{A} = \begin{bmatrix} [1+\alpha,\ 3-\alpha] & [\alpha,\ 3-2\alpha] \\ [7+5\alpha,\ 14-2\alpha] & [1+2\alpha,\ 8-5\alpha] \end{bmatrix}$$

$$= \begin{bmatrix} a_{11}^\alpha & a_{12}^\alpha \\ a_{21}^\alpha & a_{22}^\alpha \end{bmatrix}$$

Equation 3.71 may now be represented in the following way:

$$\widetilde{A}\widetilde{X} = \widetilde{\lambda}\widetilde{X}$$

$$\Rightarrow \begin{bmatrix} a_{11}^\alpha & a_{12}^\alpha \\ a_{21}^\alpha & a_{22}^\alpha \end{bmatrix}\widetilde{X} = \widetilde{\lambda}\widetilde{X}$$

TABLE 3.2

Eigenvalues and Eigenvectors for Different α Level Set

α	Eigenvectors		Eigenvalues
0	$\left\{\begin{bmatrix} -0.5593,\,0 \\ 0.8290,\,1.0000 \end{bmatrix}\right\}$	$\left\{\begin{bmatrix} -0.3027,\,0 \\ -1.0000,\,-0.9531 \end{bmatrix}\right\}$	[−1.4462, 1], [1, 12.4462]
0.1	$\left\{\begin{bmatrix} -0.5432,\,-0.1214 \\ 0.8396,\,0.9926 \end{bmatrix}\right.$	$\left.\begin{bmatrix} -0.2992,\,-0.1084 \\ -0.9941,\,-0.9542 \end{bmatrix}\right\}$	[−1.4280, 0.2825], [2.0175, 11.8280]
0.2	$\left\{\begin{bmatrix} -0.5258,\,-0.1687 \\ 0.8506,\,0.9857 \end{bmatrix}\right.$	$\left.\begin{bmatrix} -0.2955,\,-0.1446 \\ -0.9895,\,-0.9553 \end{bmatrix}\right\}$	[−1.4063, 0.0311], [2.5689, 11.2063]
0.3	$\left\{\begin{bmatrix} -0.5069,\,-0.2021 \\ 0.8620,\,0.9794 \end{bmatrix}\right.$	$\left.\begin{bmatrix} -0.2913,\,-0.1686 \\ -0.9857,\,-0.9566 \end{bmatrix}\right\}$	[−1.3808, −0.1539], [3.0539, 10.5808]
0.4	$\left\{\begin{bmatrix} -0.4865,\,-0.2280 \\ 0.8737,\,0.9737 \end{bmatrix}\right.$	$\left.\begin{bmatrix} -0.2867,\,-0.1864 \\ -0.9825,\,-0.9580 \end{bmatrix}\right\}$	[−1.3507, −0.3079], [3.5079, 9.9507]
0.5	$\left\{\begin{bmatrix} -0.4643,\,-0.2491 \\ 0.8857,\,0.9685 \end{bmatrix}\right.$	$\left.\begin{bmatrix} -0.2816,\,-0.2005 \\ -0.9797,\,-0.9595 \end{bmatrix}\right\}$	[−1.3151, −0.4437], [3.9437, 9.3151]
0.6	$\left\{\begin{bmatrix} -0.4401,\,-0.2668 \\ 0.8980,\,0.9638 \end{bmatrix}\right.$	$\left.\begin{bmatrix} -0.2758,\,-0.2119 \\ -0.9773,\,-0.9612 \end{bmatrix}\right\}$	[−1.2729, −0.5678], [4.3678, 8.6729]
0.7	$\left\{\begin{bmatrix} -0.4135,\,-0.2818 \\ 0.9105,\,0.9595 \end{bmatrix}\right.$	$\left.\begin{bmatrix} -0.2693,\,-0.2214 \\ -0.9752,\,-0.9631 \end{bmatrix}\right\}$	[−1.2228, −0.6836], [4.7836, 8.0228]
0.8	$\left\{\begin{bmatrix} -0.3844,\,-0.2948 \\ 0.9232,\,0.9556 \end{bmatrix}\right.$	$\left.\begin{bmatrix} -0.2617,\,-0.2295 \\ -0.9733,\,-0.9651 \end{bmatrix}\right\}$	[−1.1626, −0.7933], [5.1933, 7.3626]
0.9	$\left\{\begin{bmatrix} -0.3521,\,-0.3062 \\ 0.9360,\,0.9520 \end{bmatrix}\right.$	$\left.\begin{bmatrix} -0.2530,\,-0.2364 \\ -0.9716,\,-0.9675 \end{bmatrix}\right\}$	[−1.0897, −0.8985], [5.5985, 6.6897]
1	$\left\{\begin{bmatrix} -0.3162,\,-0.3162 \\ 0.9487,\,0.9487 \end{bmatrix}\right.$	$\left.\begin{bmatrix} -0.2425,\,-0.2425 \\ -0.9701,\,-0.9701 \end{bmatrix}\right\}$	[−1, −1], [6, 6]

$$\Rightarrow \left(\begin{bmatrix} a_{11}^{\alpha} - \tilde{\lambda} & a_{12}^{\alpha} \\ a_{21}^{\alpha} & a_{22}^{\alpha} - \tilde{\lambda} \end{bmatrix} \right) \tilde{X} = 0 \tag{3.74}$$

Equation 3.74 may be solved by using the traditional method and then the limit method to get uncertain eigenvalues for different membership functions (α). Using these uncertain eigenvalues, the corresponding eigenvectors are investigated. The obtained values of eigenvalues and eigenvectors for different membership functions are given in Table 3.2.

Example 3.7

We now take a generalized fuzzy eigenvalue problem (Chiao, 1998)

$$\tilde{A}\tilde{X} = \tilde{\lambda}\tilde{B}\tilde{X} \tag{3.75}$$

Here,

$$\tilde{A} = \begin{bmatrix} [10,\,15,\,20] & [5,\,6,\,7] \\ [5,\,6,\,7] & [8,\,10,\,12] \end{bmatrix} \text{ and } \tilde{B} = \begin{bmatrix} [7,\,9,\,11] & [1,\,2,\,3] \\ [2,\,3,\,4] & [6,\,8,\,10] \end{bmatrix}$$

Now, applying this algorithm, matrices $\widetilde{A}$ and $\widetilde{B}$ become

$$\widetilde{A} = \begin{bmatrix} a_{11}^{\alpha} & a_{12}^{\alpha} \\ a_{21}^{\alpha} & a_{22}^{\alpha} \end{bmatrix} \quad \text{and} \quad \widetilde{B} = \begin{bmatrix} b_{11}^{\alpha} & b_{12}^{\alpha} \\ b_{21}^{\alpha} & b_{22}^{\alpha} \end{bmatrix}.$$

Equation 3.75 may be represented in the following way:

$$\widetilde{A}\widetilde{X} = \widetilde{\lambda}\widetilde{B}\widetilde{X}$$

$$\Rightarrow \begin{bmatrix} a_{11}^{\alpha} & a_{12}^{\alpha} \\ a_{21}^{\alpha} & a_{22}^{\alpha} \end{bmatrix} \widetilde{X} = \widetilde{\lambda} \begin{bmatrix} b_{11}^{\alpha} & b_{12}^{\alpha} \\ b_{21}^{\alpha} & b_{22}^{\alpha} \end{bmatrix} \widetilde{X}$$

$$\Rightarrow \left(\begin{bmatrix} a_{11}^{\alpha} - \widetilde{\lambda}b_{11}^{\alpha} & a_{12}^{\alpha} - \widetilde{\lambda}b_{12}^{\alpha} \\ a_{21}^{\alpha} - \widetilde{\lambda}b_{21}^{\alpha} & a_{22}^{\alpha} - \widetilde{\lambda}b_{22}^{\alpha} \end{bmatrix} \right) \widetilde{X} = 0 \tag{3.76}$$

Equation 3.76 may be solved by the traditional method, and applying the limit method we get uncertain eigenvalues. Using these uncertain eigenvalues, corresponding eigenvectors are investigated. The obtained values of eigenvalues and eigenvectors for different membership functions are depicted in Table 3.3.

TABLE 3.3

Eigenvalues and Eigenvectors for Different α Level Set

α	Eigenvectors		Eigenvalues
0	$\begin{bmatrix} -1, -1 \\ -0.6461, 0.0408 \end{bmatrix}$	$\begin{bmatrix} -0.9477, -0.4619 \\ 1, 1 \end{bmatrix}$	[1.7304, 1.8124], [0.7946, 1.0754]
0.1	$\begin{bmatrix} -1, -1 \\ -0.5958, 0.0219 \end{bmatrix}$	$\begin{bmatrix} -0.9178, -0.4783 \\ 1, 1 \end{bmatrix}$	[1.7235, 1.8022], [0.8237, 1.0712]
0.2	$\begin{bmatrix} -1, -1 \\ -0.5467, 0.0018 \end{bmatrix}$	$\begin{bmatrix} -0.8881, -0.4956 \\ 1, 1 \end{bmatrix}$	[1.7183, 1.7921], [0.8506, 1.0665]
0.3	$\begin{bmatrix} -1, -1 \\ -0.4990, -0.0196 \end{bmatrix}$	$\begin{bmatrix} -0.8587, -0.5138 \\ 1, 1 \end{bmatrix}$	[1.7148, 1.7824], [0.8754, 1.0612]
0.4	$\begin{bmatrix} -1, -1 \\ -0.4529, -0.0425 \end{bmatrix}$	$\begin{bmatrix} -0.8296, -0.5329 \\ 1, 1 \end{bmatrix}$	[1.7128, 1.7730], [0.8983, 1.0551]
0.5	$\begin{bmatrix} -1, -1 \\ -0.4085, -0.0669 \end{bmatrix}$	$\begin{bmatrix} -0.8010, -0.5528 \\ 1, 1 \end{bmatrix}$	[1.7122, 1.7640], [0.9194, 1.0483]
0.6	$\begin{bmatrix} -1, -1 \\ -0.3659, -0.0929 \end{bmatrix}$	$\begin{bmatrix} -0.7728, -0.5738 \\ 1, 1 \end{bmatrix}$	[1.7129, 1.7554], [0.9387, 1.0407]
0.7	$\begin{bmatrix} -1, -1 \\ -0.3251, -0.1205 \end{bmatrix}$	$\begin{bmatrix} -0.7452, -0.5956 \\ 1, 1 \end{bmatrix}$	[1.7149, 1.7474], [0.9563, 1.0321]
0.8	$\begin{bmatrix} -1, -1 \\ -0.2862, -0.1499 \end{bmatrix}$	$\begin{bmatrix} -0.7183, -0.6184 \\ 1, 1 \end{bmatrix}$	[1.7180, 1.7400], [0.9723, 1.0225]
0.9	$\begin{bmatrix} -1, -1 \\ -0.2493, -0.1812 \end{bmatrix}$	$\begin{bmatrix} -0.6921, -0.6421 \\ 1, 1 \end{bmatrix}$	[1.7222, 1.7332], [0.9868, 1.0119]
1	$\begin{bmatrix} -1, -1 \\ -0.2143, -0.2143 \end{bmatrix}$	$\begin{bmatrix} -0.6667, -0.6667 \\ 1, 1 \end{bmatrix}$	[1.7273, 1.7273], [1, 1]

TABLE 3.4

Comparisons of Eigenvalues of Results Obtained in Example 3.7 Using Vertex Method

Eigenvalues	
Vertex Method	**Limit Method**
[1.7273, 1.7304, 1.8124] and [0.7946, 1, 1.0754]	[1.7122, 1.7273, 1.8124] and [0.7946, 1, 1.0754]

Furthermore, the investigated eigenvalues of Example 3.7 are compared with the results obtained by the vertex method. The comparisons are presented in Table 3.4. Here, we observe that the limit method is better compared to the vertex method in the sense of the distribution of uncertain eigenvalues. The width of the eigenvalues obtained in the limit method is more in comparison with the vertex method, which shows the possibility of more numbers of eigenvalues.

Example 3.8

Finally, we consider the neutron diffusion problem for a one-group square homogeneous reactor, which is represented in Figure 3.3.

The governing differential equation for the bare homogeneous reactor (Chakraverty and Nayak 2013) with fuzzy parameters may be written as

$$\widetilde{D}\nabla^2\tilde{\phi} + \tilde{S} = \tilde{\Sigma}_a\tilde{\phi} \tag{3.77}$$

The boundary condition is $\tilde{\phi}(x, \pm 1) = 0 = \tilde{\phi}(\pm 1, y)$, where $\widetilde{D}$, $\tilde{\phi}$, $\tilde{S}$, $\tilde{\Sigma}_a$ are fuzzy diffusion coefficient, neutron flux, source and absorption coefficient, respectively, and their corresponding values are presented in Table 3.5.

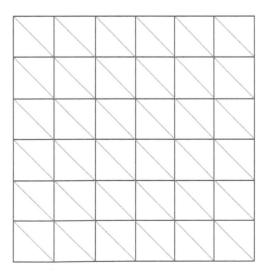

FIGURE 3.3

72 element discretization of the domain.

TABLE 3.5

Fuzzy Values of Involved Parameters

Parameters	TFN
$\widetilde{D}$	$[0.5+0.5\alpha,\ \ 1.5-0.5\alpha]$
$\widetilde{\Sigma}_a$	$[0.5+0.5\alpha,\ \ 1.5-0.5\alpha]$
$\widetilde{S}$	$[1, 1]$

TABLE 3.6

Largest and Smallest Fuzzy Eigenvalues

	Smallest	Largest
Eigenvalues	$[0.0527, 1.75900, 158.3]$	$[2.0528, 68.4288, 6158.5]$

In this problem, the domain is discretized into 72 triangular elements. Then, using the finite element procedure, the governing differential Equation 3.77 is converted into the following fuzzy algebraic equation:

$$\left[\widetilde{K}\right]\left\{\widetilde{\phi}\right\} = \left\{\widetilde{Q}\right\} \tag{3.78}$$

where

$\left[\widetilde{K}\right]$ is the assembled fuzzy stiffness matrix corresponding to leakage and absorption terms

$\left\{\widetilde{Q}\right\}$ is the assembled fuzzy force vector for the source term

Here, Equation 3.78 represents the eigenvalue problem and is solved by using the proposed algorithm. The results are given for the smallest and largest eigenvalues in Table 3.6.

Bibliography

Abbasbandy, S. and Alvi, M. 2005. A method for solving fuzzy linear systems. *Iranian Journal of Fuzzy Systems* 2:37–43.

Allahviranloo, T., Kermani, M. A. and Shafiee, M. 2008. Revised solution of an overdetermined fuzzy linear system of equations. *International Journal of Computational Cognition* 6(3):66–70.

Black, F. and Scholes, M. 1973. The pricing of options and corporate liabilities. *Journal of Political Economy* 81:637–654.

Chakraverty, S. and Nayak, S. 2013. Non probabilistic solution of uncertain neutron diffusion equation for imprecisely defined homogeneous bare reactor. *Annals of Nuclear Energy* 62:251–259.

Chiao, K. P. 1998. Generalized fuzzy eigenvalue problems. *Tamsui Oxford Journal of Mathematical Sciences* 14:31–37.

Dempster, A. P. 1967. Upper and lower probabilities induced by multivalued mapping. *Annals of Mathematical Statistics* 38(2):325–339.

Ferson, S. and Ginzburg, L. R. 1996. Different methods are needed to propagate ignorance and variability. *Reliability Engineering & System Safety* 54:133–144.

Friedman, M., Ming, M. and Kandel, A. 1998. Fuzzy linear systems. *Fuzzy Sets and Systems* 96:201–209.

Gelman, A. 2006. Prior distributions for variance parameters in hierarchical models (comment on article by Browne and Draper). *Bayesian Analysis* 1(3):515–534.

Ghanem, R. and Spanos, P. 2003. *Stochastic Finite Elements: A Spectral Approach*. Courier Dover Publications, USA.

Higham, D. J. and Kloeden, P. 2005. Numerical methods for nonlinear stochastic differential equations with jumps. *Numerische Mathematik* 101:101–119.

Li, J., Li, W. and Kong, X. 2010. A new algorithm model for solving fuzzy linear systems. *Southeast Asian Bulletin of Mathematics* 34:121–132.

Matinfar, M., Nasseri, S. H. and Sohrabi, M. 2008. Solving fuzzy linear system of equations by using householder decomposition method. *Applied Mathematical Sciences* 52(2):2569–2575.

Moore, R. E., Kearfott, R. B. and Cloud, M. J. 2014. *Introduction to Interval Analysis*. SIAM Press, Philadelphia, PA.

Nayak, S. and Chakraverty, S. 2013. Non-probabilistic approach to investigate uncertain conjugate heat transfer in an imprecisely defined plate. *International Journal of Heat and Mass Transfer* 67:445–454.

Nayak, S. and Chakraverty, S. 2015. Numerical solution of uncertain neutron diffusion equation for imprecisely defined homogeneous triangular bare reactor. *Sadhana* 40:2095–2109.

Nayak, S. and Chakraverty, S. 2016. Numerical solution of stochastic point kinetic neutron diffusion equation with fuzzy parameters. *Nuclear Technology* 193(3):444–456.

Neumaier, A. 1990. *Interval Methods for Systems of Equations*. Cambridge University Press, New York.

Sauer, T. 2012. *Numerical Solution of Stochastic Differential Equations in Finance*. Springer, USA.

Schöbi, R. and Sudret, B. 2015a. Propagation of uncertainties modelled by parametric P-boxes using sparse polynomial chaos expansions. In: *Twelfth International Conference on Applications of Statistics and Probability in Civil Engineering, ICASP12*, Vancouver, British Columbia, Canada.

Schöbi, R. and Sudret, B. 2015b. Imprecise structural reliability analysis using PC-kriging. In: *Safety and Reliability of Complex Engineered Systems. Proceedings of the 25th European Safety and Reliability Conference (ESREL'2015)*, CRC Press, Zurich, Switzerland.

Senthilkumar, P. and Rajendran, G. 2011. New approach to solve symmetric fully fuzzy linear systems. *Sadhana* 36(6):933–940.

Sevastjanov, P. and Dymova, L. 2009. A new method for solving interval and fuzzy equations: Linear case. *Information Sciences* 6:263–274.

Sudret, B. 2014. Polynomial chaos expansions and stochastic finite element methods. *Risk and Reliability in Geotechnical Engineering*, Phoon, K. and Ching, J. (Eds.), CRC Press, 265–300.

4

Uncertain Neutron Diffusion

Generally, every system possesses uncertainties caused by measurement errors, vague data and boundary conditions. Accordingly, the involved coefficients, parameters and constants become uncertain. Neutron diffusion is the backbone of a nuclear reactor. The scattering of a neutron occurs when it undergoes diffusion, which involves uncertainty due to the reactor parameters, measurement errors and boundary conditions. These uncertainties play an important role in the problems related to reactors, which should be understood well to help in a better design. In this chapter, we discuss the factors which may be considered as uncertain in a reactor system and then incorporate the corresponding idea of modelling.

4.1 Uncertain Factors Involved in Neutron Diffusion Theory

The neutron collision inside a reactor depends upon the geometry of the reactor, diffusion coefficient, absorption coefficient, etc. In general, these parameters are not crisp and hence we get an uncertain neutron diffusion equation. Here, these uncertain parameters are taken as fuzzy. To investigate the uncertain spectrum of neutron flux distribution, we formulate the fuzzy finite element method (FFEM) with the linear triangular fuzzy element discretizing the domain.

When neutrons undergo diffusion, they suffer scattering collisions with the nuclei, assumed to be initially stationary, and as a result, a typical neutron trajectory divides into a number of short path elements. These are scattering free paths. The average of these is the mean free path. When a large number of neutrons are considered, there is a net movement of neutrons from regions of higher to those of lower concentration. As the path of a neutron after scattering may not be known exactly, we may take the cross section and transport the mean free path as fuzzy. As a result, the diffusion coefficient lies in an uncertain region and becomes fuzzy. Similarly, the absorption coefficient may also be taken as fuzzy.

4.1.1 Diffusion Coefficient and Absorption Coefficient

Scattering or absorption reaction occurs when neutrons interact with matter. A change in the energy and direction of motion of a neutron happens when scattering takes place, but it cannot directly cause the disappearance of a free neutron, whereas absorption results in the disappearance of free neutrons. Hence, the nuclear reaction takes place with fission or the formation of a new nucleus and another particle or particles such as protons, alpha particles and gamma photons (Ghoshal 2010).

The effect of the scattering angular distribution of the motion of a neutron depends on the path and deflection. After each collision, the particle travels a scattering mean free path λ_s and is deflected by an angle θ, which is uncertain in general. When the deflection angle θ is taken as fuzzy, we have the projected distance travelled after the first collision along the z-axis as

$$\tilde{Z}_1 = \lambda_s \cos \tilde{\theta}_1$$

The average value of $\tilde{Z}_1$ is

$$\bar{\tilde{Z}}_1 = \lambda_s \langle \cos \tilde{\theta}_1 \rangle = \lambda_s \bar{\tilde{\mu}} \tag{4.1}$$

where $\bar{\tilde{\mu}} = (2/3\tilde{A})$; $\tilde{A}$ is the mass number of the uncertain nuclei in the scattering medium. The distance travelled after the second collision is

$$\tilde{Z}_2 = \lambda_s \langle \cos \tilde{\theta}_1 \rangle \langle \cos \tilde{\theta}_2 \rangle$$

The average value of $\tilde{Z}_2$ is

$$\bar{\tilde{Z}}_2 = \lambda_s \langle \cos \tilde{\theta}_1 \rangle \langle \cos \tilde{\theta}_2 \rangle \approx \lambda_s \langle \cos \tilde{\theta}_1 \rangle^2 = \lambda_s \bar{\tilde{\mu}}^2 \tag{4.2}$$

In general, for n collisions

$$\bar{\tilde{Z}}_n = \lambda_s \bar{\tilde{\mu}}^n, \quad \text{for } n = 0, 1, 2, \ldots \tag{4.3}$$

Since $\bar{\tilde{Z}}_n \to 0$ as $n \to \infty$, this implies that the neutron could be scattered in either direction with equal probability at its next collision. This means that the neutron forgets its original direction of motion after a succession of collisions, which is characteristic of Markov chains, and having been carried a distance of a transport mean free path:

$$\lambda_{tr} = \bar{\tilde{Z}}_0 + \bar{\tilde{Z}}_1 + \cdots = \sum_{n=0}^{\infty} \lambda_s \bar{\tilde{\mu}}^n$$

$$= \lambda_s \left(1 + \bar{\tilde{\mu}} + \bar{\tilde{\mu}}^2 + \cdots \right)$$

$$= \frac{\lambda_s}{1 - \bar{\tilde{\mu}}}, \quad \forall \bar{\tilde{\mu}} < 1 \tag{4.4}$$

The transport cross section is defined as

$$\Sigma_{tr} = \frac{1}{\lambda_{tr}} = \frac{1 - \bar{\tilde{\mu}}}{\lambda_s} = \Sigma_s \left(1 - \bar{\tilde{\mu}} \right) \tag{4.5}$$

If absorption is present, we generalize the definition of transport cross section to be

$$\Sigma_{tr} = \Sigma_a + \Sigma_s (1 - \bar{\tilde{\mu}})$$

$$= (\Sigma_a + \Sigma_s) - \Sigma_s \bar{\tilde{\mu}} = \Sigma_t - \Sigma_s \bar{\tilde{\mu}} \tag{4.6}$$

where Σ_t is the total macroscopic cross section.

Hence, we may conclude that Σ_{tr} is uncertain. As such, both the diffusion and absorption coefficients become uncertain.

Due to the uncertain diffusion and absorption coefficients, the one-group uncertain neutron diffusion equation for bare reactor becomes

$$\tilde{D}\nabla^2\phi + S = \tilde{\Sigma}_a\phi, \tag{4.7}$$

where '~' represents fuzziness.

Similarly, the general form of the neutron diffusion equation (multigroup neutron diffusion) (Glasstone and Sesonke 2004) in terms of fuzzy may be written as

$$\frac{d}{dx}\left[\tilde{D}_g(x)\frac{d\tilde{\phi}_g(x)}{dx}\right] - \sum_g^t(x)\tilde{\phi}_g(x) + \sum_{g'=1}^{G}\sum_{g'\to g}^{s}(x)\tilde{\phi}_{g'}(x) + \tilde{F}_g(x) = 0 \tag{4.8}$$

where $g = 1, 2, ..., G$.

In Equation 4.8, the first, second, third and fourth terms represent leakage of neutrons from group g, the total rate of neutron interaction, scattering of neutrons from other groups into group g and the rate at which neutrons are produced in the group, respectively. Furthermore, these terminologies are defined in Chapter 7.

These fuzzy models can be investigated by using the interval/FFEM. The finite element method (FEM) is a numerical technique, which may be used to obtain approximate solutions for ordinary or partial differential equations. In particular, this method may be used to transform partial differential equations into algebraic equations, which may then be easily solved.

The steps (Gerald and Wheatley 2003, Hutton 2005) involved in the FEM for crisp parameters are

1. Finding the functional that corresponds to the partial differential equation. This is well known for a large class of problems. The functional may be developed by using the Galarkin technique of weighted residue methods.

2. Subdividing the region into subregions or elements. The elements must span the entire region and approximate the boundary relatively closely.

3. Writing an interpolating a relation that gives values for the dependent variable within an element based on the values at the nodes.

4. Substituting the interpolating relation into the functional and setting the partial derivatives of the functional with respect to each scalar coefficient (from Galarkin technique) to zero.

5. Combining the element equations of step 4. This way we get a system of equations, which we adjust for the boundary conditions of the problem, then solve it.

In this procedure, when crisp parameters are replaced by interval/fuzzy, the problem becomes complicated and then the FEM is to be developed intelligently. As such, the developed interval/FFEM has been discussed here.

4.1.2 Fuzzy Finite Element Method

If the parameters as well as initial and boundary conditions are uncertain, then the governing differential equations become uncertain. Accordingly, the uncertainties are considered as intervals/fuzzy (Hanss 2005), and interval/FFEM is developed. Considering field variables and the involved parameters as interval/fuzzy, we get interval/fuzzy element properties in terms of interval/fuzzy matrices. Then, those matrices are assembled and the global matrix is obtained. Furthermore, boundary conditions are imposed, which may also be interval/fuzzy. From the global matrices and boundary conditions, we get a system of algebraic equations that are either a system of simultaneous equations or eigenvalue problems. It may be noted that due to the uncertainties (i.e. interval/fuzzy), we now have the interval/fuzzy system of simultaneous equations or eigenvalue problems. Here, a system of interval/fuzzy algebraic equations are investigated by using the interval/fuzzy arithmetic (limit method).

Figure 4.1 shows a schematic diagram of the fuzzy finite element procedure, which gives the basic idea to encrypt the process of the FFEM. We have modified the usual interval/fuzzy arithmetic in the FEM. It involves three steps: input, hidden layer and output. In the input step, we have considered uncertain parameters and field variables. These uncertain parameters are taken as fuzzy. In the hidden layer step, element properties are obtained by using various fuzzy parameters. The element properties and fuzzy stiffness matrices are assembled and finally the global stiffness matrices are developed. Further, initial and boundary conditions are imposed and the transformed fuzzy system of equations is solved through the limit technique (Nayak and Chakraverty 2013), and various sub-steps are executed inside the hidden layer. Finally, we get uncertain fuzzy solutions as the output, which may be different in type and nature corresponding to the input fuzzy parameters. Here, in Figure 4.1, we have considered triangular fuzzy numbers as input parameters. The alpha (α)-level representation of two fuzzy sets $\tilde{X}$ and $\tilde{Y}$ with their triangular membership functions for the fuzzy arithmetic operation (Nayak and Chakraverty 2013) is shown in Figure 4.1. The deterministic value is obtained for α_4 level of fuzzy sets, whereas for α_1, α_2 and α_3 levels, we get different interval values. If we consider the value of α to be zero, then the deterministic interval lies on the x-axis. The output may be generated by considering all possible combinations of the alpha (α) levels.

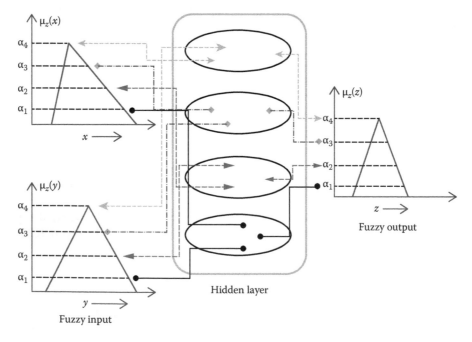

FIGURE 4.1
Model diagram of the fuzzy finite element procedure.

4.1.3 Uncertainty Caused by the Stochastic Process

The interactions and scattering neutrons in reactors follow stochastic processes. In neutron diffusion, the neutron populations show randomness. As such, a birth–death stochastic process is found in these systems. Hence, the governing differential equation for neutron diffusion is converted into stochastic.

The concept of stochastic differential equation (SDE) was initiated by the great philosopher Einstein in 1905 (Sauer 2012). A mathematical connection between the microscopic random motion of particles and the macroscopic diffusion equation was presented. Later, it was seen that the SDE model plays a prominent role in a range of application areas such as mathematics, physics, chemistry, mechanics, biology, microelectronics, economics and finance. Earlier, SDEs were solved using the Ito integral as an exact method, which is discussed in Malinowski and Michta (2011). But using the exact method proved difficult to study nontrivial problems and hence approximation methods came to be used. In 1982, Rumelin (1982) defined general Runge–Kutta approximations for the solution of SDEs and an explicit form of the correction term has been given. Kloeden and Platen (1992) discussed about the numerical solutions of the SDE. Discrete time strong and weak approximation methods were used by Platen (1999) to investigate the solution of SDEs. Next, Higham (2001) gave a major contribution to solve the approximate solutions of SDEs. Furthermore, Higham and Kloeden (2005) investigated nonlinear SDEs numerically. They presented two implicit methods for the Ito SDEs with Poisson-driven jumps. The first method is a split-step extension of the backward Euler method, and the second method arises from the introduction of a compensated, martingale, form of the Poisson process. Hayes and Allen (2005) solved the stochastic point kinetic reactor problem. They modelled the point stochastic reactor problem into the ordinary time-dependent SDE and studied the stochastic behaviour of the neutron flux.

4.1.4 Hybrid Uncertainties

It may be noted from the literature review that, previously, authors investigated the SDEs which contained crisp parameters. As mentioned earlier, in general, the involved parameters may not be crisp, but rather uncertain. As such, Kim (2005) considered fuzzy sets space for real line was considered by Kim (2005) and the existence and uniqueness of the solution are obtained. The solution is investigated by taking particular conditions, which are imposed on the structure of the integrated fuzzy stochastic processes such that a maximal inequality for the fuzzy stochastic Ito integral holds. Next, Ogura (2008) proposed an approach to solve the fuzzy stochastic differential equation (FSDE), which does not contain any notion of the fuzzy stochastic Ito integral, and the method was based on the selection of sets. Further, Malinowski and Michta (2011) presented the existence and uniqueness of solutions to the FSDEs driven by the Brownian motion, and the continuous dependence on initial condition and stability properties are studied.

4.2 Modelling of the Uncertain Neutron Diffusion Equation

As discussed earlier, because of the uncertainties caused by diffusion and absorption coefficients, the one- and multigroup neutron diffusion equation becomes uncertain. If the uncertainties are in terms of interval, then these equations are called interval one- or multigroup neutron diffusion equations. Whereas if the uncertainties are fuzzy then these equations become fuzzy one- or multigroup neutron diffusion equations. In time-dependent neutron diffusion problems, that is point kinetic neutron diffusion, equations possess uncertainties, which result in uncertain neutron populations. Hence, it follows that stochastic (such as birth–death) processes and the coefficients are uncertain (fuzzy) in nature. As such, both the stochastic and fuzzy theory may be combined to model the uncertain neutron diffusion equation. The stochastic point kinetic neutron diffusion equation is

$$\frac{\partial N}{\partial t} = Dv\nabla^2 N - \left(\Sigma_a - \Sigma_f\right)vN + \left[(1-\beta)k_\infty\Sigma_a - \Sigma_f\right]vN + \sum_i \lambda_i C_i + S_0 \quad \text{for } i = 1, 2, \cdots, m \quad (4.9)$$

$$\frac{\partial C_i}{\partial t} = \beta_i k_\infty \Sigma_a vN - \lambda_i C_i \quad \text{for } i = 1, 2, \cdots, m \quad (4.10)$$

where
$N = N(r, t)$ is the neutron density at position r at time t
v is the velocity of the neutron
Σ_f is the neutron fission cross section
D is the diffusion coefficient
Σ_a is the absorption coefficient
$\beta = \sum_{i=1}^{m} \beta_i$ is the delayed neutron fraction

$1 - \beta$ is the prompt neutron fraction
k_∞ is the infinite medium neutron reproduction factor
λ_i is the delay constant
$C_i = C_i(r, t)$ the density of the ith type of precursors at position r at time t
S_0 is the extraneous neutron source
$Dv\nabla^2 N$ is the diffusion term of the neutrons
$(\Sigma_a - \Sigma_f)$ is the capture cross section
$[(1 - \beta)k_\infty \Sigma_a - \Sigma_f]vN$ is the prompt neutron contribution to the source
$\sum_{i=1}^{m} \lambda_i C_i$ is the rate of transformations from the neutron precursors to the neutron population

Let us consider uncertain parameters, viz. delayed neutron fraction, source and initial condition as fuzzy in the SDEs (Equations 4.9 and 4.10). Then we have the following FSDEs:

$$\frac{\partial \tilde{N}}{\partial t} = Dv\nabla^2 \tilde{N} - \left(\Sigma_a - \Sigma_f\right)v\tilde{N} + \left[(1-\tilde{\beta})k_\infty\Sigma_a - \Sigma_f\right]v\tilde{N} + \sum_i \lambda_i C_i + S_0 \quad (4.11)$$

$$\frac{\partial \tilde{C}_i}{\partial t} = \tilde{\beta}_i k_\infty \Sigma_a v\tilde{N} - \lambda_i \tilde{C}_i \quad \text{for } i = 1, 2, \ldots, m \quad (4.12)$$

where '~' represents fuzzy numbers.

Using both the birth–death stochastic process and fuzzy theory, Equations 4.11 and 4.12 will be transformed into the following algebraic equations (only one precursor taken):

$$\frac{d}{dt}\begin{bmatrix} \tilde{n} \\ \tilde{c}_1 \end{bmatrix} = \tilde{\tilde{A}}\begin{bmatrix} \tilde{n} \\ \tilde{c}_1 \end{bmatrix} + \begin{bmatrix} \tilde{q} \\ 0 \end{bmatrix} + \tilde{\tilde{B}}^{1/2}\frac{d\vec{W}}{dt} \tag{4.13}$$

where

$$\tilde{\tilde{A}} = \begin{bmatrix} \dfrac{\rho-\tilde{\beta}}{l} & \lambda_1 \\[2ex] \dfrac{\tilde{\beta}_1}{l} & -\lambda_1 \end{bmatrix}$$

$$\tilde{\tilde{B}} = \begin{bmatrix} \gamma n + \lambda_1 c_1 + \tilde{q} & \dfrac{\tilde{\beta}_1}{l}\left(\left(1-\tilde{\beta}\right)v-1\right)n-\lambda_1 c_1 \\[2ex] \dfrac{\tilde{\beta}_1}{l}\left(\left(1-\tilde{\beta}\right)v-1\right)n-\lambda_1 c_1 & \dfrac{\tilde{\beta}_1^2\,v}{l}n+\lambda_1 c_1 \end{bmatrix}, \quad \gamma = \frac{-1-\rho+2\tilde{\beta}+\left(1-\tilde{\beta}\right)^2 v}{l}$$

$$\vec{W} = \vec{W}(t) = \begin{bmatrix} W_1(t) \\ W_2(t) \end{bmatrix}, \text{ where } W_1(t) \text{ and } W_2(t) \text{ are Wiener processes}$$

Generalizing Equations 4.11 and 4.12, the stochastic point kinetic equation for m precursors becomes

$$\frac{d}{dt}\begin{bmatrix} \tilde{n} \\ \tilde{c}_1 \\ \tilde{c}_2 \\ \vdots \\ \tilde{c}_m \end{bmatrix} = \tilde{\tilde{A}}\begin{bmatrix} \tilde{n} \\ \tilde{c}_1 \\ \tilde{c}_2 \\ \vdots \\ \tilde{c}_m \end{bmatrix} + \begin{bmatrix} \tilde{q} \\ 0 \\ 0 \\ \vdots \\ 0 \end{bmatrix} + \tilde{\tilde{B}}^{1/2}\frac{d\vec{W}}{dt} \tag{4.14}$$

where

$$\tilde{\tilde{A}} = \begin{bmatrix} \dfrac{\rho-\tilde{\beta}}{l} & \lambda_1 & \lambda_2 & \cdots & \lambda_m \\[2ex] \dfrac{\tilde{\beta}_1}{l} & -\lambda_1 & 0 & \cdots & 0 \\[2ex] \dfrac{\tilde{\beta}_2}{l} & 0 & -\lambda_2 & \ddots & \vdots \\[2ex] \vdots & \vdots & \ddots & \ddots & 0 \\[2ex] \dfrac{\tilde{\beta}_m}{l} & 0 & \cdots & 0 & -\lambda_m \end{bmatrix}$$

$$\tilde{\tilde{B}} = \begin{bmatrix} \tilde{\zeta} & \tilde{a}_1 & \tilde{a}_2 & \cdots & \tilde{a}_m \\ \tilde{a}_1 & \tilde{r}_1 & \tilde{b}_{2,3} & \cdots & \tilde{b}_{2,m} \\ \tilde{a}_2 & 0 & \tilde{r}_2 & \ddots & \vdots \\ \vdots & \vdots & \ddots & \ddots & \tilde{b}_{m-1,m} \\ \tilde{a}_m & \tilde{b}_{m,2} & \cdots & \tilde{b}_{m,m-1} & \tilde{r}_m \end{bmatrix}$$

Here, $\tilde{\zeta} = \gamma n + \sum_{i=1}^{m} \lambda_i c_i + \tilde{q}$, $\tilde{a}_i = \frac{\tilde{\beta}_i}{l}\left(\left(1-\tilde{\beta}\right)v - 1\right)n - \lambda_i c_i$, $\tilde{b}_{i,j} = \frac{\tilde{\beta}_{i-1}\tilde{\beta}_{j-1}v}{l}n$ and $\tilde{r}_i = \frac{\tilde{\beta}_i^2 v}{l}n + \lambda_i c_i$.

Equation 4.14 may be written in compact form as

$$\frac{d}{dt}\tilde{x} = \tilde{\tilde{A}}\tilde{x} + \begin{bmatrix} \tilde{q} & 0 & \cdots & 0 \end{bmatrix}^T + \tilde{\tilde{B}}^{1/2}\frac{d\vec{W}}{dt} \tag{4.15}$$

where $\tilde{x} = \begin{bmatrix} \tilde{n} & \tilde{c}_1 & \tilde{c}_2 & \cdots & \tilde{c}_m \end{bmatrix}^T$.

Bibliography

Alefeld, G. and Mayer, G. 2000. Interval analysis: Theory and applications. *Journal of Computational and Applied Mathematics* 121:421–464.

Allahviranloo, T., Mikaeilvand, N., Kiani, N. A. and Shabestari, R. M. 2008. Signed decomposition of fully fuzzy linear systems. *Applications and Applied Mathematics* 3(1): 77–88.

Casasnovas, J. and Riera, J. V. 2007. Maximum and minimum of discrete fuzzy numbers. *Frontiers in Artificial Intelligence and Applications* 163:273–280.

Chakraverty, S. and Nayak, S. 2012. Fuzzy finite element method for solving uncertain heat conduction problems. *Coupled System Mechanics* 1(4):345–360.

Chakraverty, S. and Nayak, S. 2013. Non probabilistic solution of uncertain neutron diffusion equation for imprecisely defined homogeneous bare reactor. *Annals of Nuclear Energy* 62:251–259.

Cloud, M. J., Moore, R. E. and Kearfott, R. B. 2009. *Introduction to Interval Analysis.* Society for Industrial and Applied Mathematics, Philadelphia, PA.

Dong, W. and Shah, H. 1987. Vertex method for computing functions of fuzzy variables. *Fuzzy Sets and Systems* 24:65–78.

Dong, W. M. and Wong, F. S. 1987. Fuzzy weighted average and implementation of the extension principle. *Fuzzy Sets and Systems* 21:183–199.

Dubois, D. and Prade, H. 1980. *Theory and Application, Fuzzy Sets and Systems.* Academic Press, Waltham, MA.

Gao, L. S. 1999. The fuzzy arithmetic mean. *Fuzzy Sets and Systems* 107(3):335–348.

Gerald, C. F. and Wheatley, P. O. 2003. *Applied Numerical Analysis*, 7th edn., Pearson.

Ghoshal, S. N. 2010. *Nuclear Physics.* S. Chand & Co. Ltd., New Delhi, India.

Glasstone, S. and Sesonke, A. 2004. *Nuclear Reactor Engineering*, CBS Publishers.

Hanss, M. 2005. *Applied Fuzzy Arithmetic: An Introduction with Engineering Applications.* Springer, USA.

Hayes, J. G. and Allen, E. J. 2005. Stochastic point kinetic equations in nuclear reactor dynamics. *Annals of Nuclear Energy* 32:572–587.

Higham, D. J. 2001. An algorithmic introduction to numerical simulation of stochastic differential equations. *SIAM Review* 34:525–546.

Higham, D. J. and Kloeden, P. 2005. Numerical methods for nonlinear stochastic differential equations with jumps. *Numerische Mathematik* 101:101–119.

Hutton, D. V. 2005. *Fundamental of Finite Element Analysis.* Tata McGraw-Hill, New York.

Kim, J. H. 2005. On fuzzy stochastic differential equations, *Journal of Korean Mathematical Society*, 42: 153–169.

Kloeden, P. and Platen, E. 1992. *Numerical Solution of Stochastic Differential Equations.* Springer, Berlin, Germany.

Malinowski, M. T. and Michta, M. 2011. Stochastic fuzzy differential equations with an application. *Kybernetika* 47(1):123–143.

Matinfar, M., Nasseri, H. S. and Sohrabi, M. 2008. Solving fuzzy linear system of equations by using householder decomposition method. *Applied Mathematical Sciences* 2(52):2569–2575.

Nayak, S. and Chakraverty, S. 2013. Non-probabilistic approach to investigate uncertain conjugate heat transfer in an imprecisely defined plate. *International Journal of Heat and Mass Transfer* 67:445–454.

Ogura, Y. 2008. *On Stochastic Differential Equations with Fuzzy Set Coefficients. Soft Methods for Handling Variability and Imprecision.* Springer, Berlin, Germany.

Oksendal, B. 2003. *Stochastic Differential Equations: An Introduction with Applications.* Springer-Verlag, Heidelberg, Germany.

Platen, E. 1999. An introduction to numerical methods for stochastic differential equations. *Acta Numerica* 8:197–246.

Rumelin, W. 1982. Numerical treatment of stochastic differential equations, *SIAM Journal of Numerical Analysis*, 19: 604–613.

Sauer, T. 2012. *Numerical Solution of Stochastic Differential Equations in Finance.* Springer, USA.

Zadeh, L. A. 1965. Fuzzy sets. *Information and Control* 8:338–353.

Zimmermann, H. J. 1991. *Fuzzy Sets Theory and Its Applications.* Kluwer Academic Press, Dordrecht, the Netherlands.

5

One-Group Model

In this chapter, various methods have been discussed to investigate the one-group model of neutron diffusion problems. These problems may be solved by analytical methods for simple cases only. When analytical methods may not be used to solve, then we may utilize numerical methods such as the finite difference method (FDM) and the finite element method (FEM).

5.1 Analytical Method

In physical problems, we may have some specified conditions known as the initial or boundary conditions, which are to be considered with the governing differential equation. The solution of the differential equation satisfies these initial boundary conditions. Hence, the differential equation, together with these initial or boundary conditions, forms an initial value or boundary value problem.

The problems which are governed by ordinary differential equations are solved in two steps. First, the general solution is to be found and then we determine the arbitrary constants from the initial values. But the same process is not applicable to problems involving partial differential equations. In partial differential equations, the general solution contains arbitrary functions, which are difficult to be adjusted to satisfy the given boundary conditions. In the following sections, a few methods are discussed for the boundary value problems (linear partial differential equations).

5.1.1 Separation of Variable Method

This method is used to convert the partial differential equations into ordinary differential equations. It involves a solution, which splits up into a product of functions, where each one contains only one variable. The steps to solve the differential equation are as follows:

1. By applying this method, we shall obtain two or more ordinary differential equations depending on the governing differential equations.
2. We shall determine solutions of those ordinary differential equations that satisfy the boundary conditions provided with the problems.
3. Using the Fourier series, we shall compose those solutions of the ordinary differential equations and get the final solution of the governing differential equations.

5.1.2 One-Dimensional Wave Equation

Let us consider the 1D wave equation (Kreysig 2010) with the boundary conditions as

$$\frac{\partial^2 u}{\partial t^2} = c^2 \frac{\partial^2 u}{\partial x^2} \tag{5.1}$$

where $c^2 = T/\rho$. Here, T and ρ are the tension and mass of the deflected string.

5.1.2.1 Boundary Conditions

Here, $u(x, t)$ is the deflection of the string, and it is fixed at the ends $x = 0$ and $x = L$. Hence, we have

$$u(0, t) = 0, \quad u(L, t) = 0, \quad \forall t \tag{5.2}$$

Denoting the initial deflection by the function $f(x)$ and the initial velocity by $g(x)$, we obtain

$$u(x, 0) = f(x) \tag{5.3}$$

and

$$\left. \frac{\partial u}{\partial t} \right|_{t=0} = g(x). \tag{5.4}$$

We determine the wave Equation 5.1 of the following form:

$$u(x, t) = F(x)G(t) \tag{5.5}$$

which is a product of two functions F and G depending on the variables x and t, respectively. By differentiating Equation 5.5, we get

$$\frac{\partial^2 u}{\partial t^2} = F\ddot{G} \tag{5.6}$$

and

$$\frac{\partial^2 u}{\partial x^2} = F''G \tag{5.7}$$

where
   $(\cdot)$ denotes the derivative with respect to t
   primes denotes the derivative with respect to x

Using Equations 5.6 and 5.7, Equation 5.1 becomes

$$F\ddot{G} = c^2 F''G \tag{5.8}$$

Equation 5.8 is transformed into

$$\frac{\ddot{G}}{c^2 G} = \frac{F''}{F}.$$

(5.9)

Then, Equation 5.9 may be written as

$$\frac{\ddot{G}}{c^2 G} = \frac{F''}{F} = k$$

(5.10)

where k is the arbitrary constant.

Equation 5.10 will give the following two ordinary linear differential equations, that is:

$$F'' - kF = 0$$

(5.11)

and

$$\ddot{G} - c^2 k G = 0.$$

(5.12)

Choosing $k = -p^2$, the general solution of Equation 5.11 becomes

$$F = A \cos px + B \sin px$$

(5.13)

We know

$$u(0, t) = F(0)G(t) = 0, \quad u(L, t) = F(L)G(t) = 0, \quad \forall t.$$

If $G = 0$, then $u = 0$ which has no interest. Hence, $G \neq 0$.

If $F = 0$, then we have the following:

$$F(0) = 0 \quad \text{and} \quad F(L) = 0.$$

(5.14)

Equations 5.14 and 5.13 give

$$F(0) = A = 0.$$

Again, we have

$$F(L) = 0 \Rightarrow B \sin pL = 0.$$

We must take $B \neq 0$, otherwise $F = 0$. Hence, $\sin pL = 0$. Thus,

$$pL = n\pi \Rightarrow p = \frac{n\pi}{L}$$

(5.15)

where n is an integer.

Taking $B = 1$, we obtain infinitely many solutions $F(x) = F_n(x)$, where

$$F_n(x) = \sin\frac{n\pi}{L}x, \quad n = 1, 2, \ldots \tag{5.16}$$

Solving Equation 5.12, we get

$$\ddot{G} + \lambda_n^2 G = 0 \tag{5.17}$$

where

$$\lambda_n^2 = -c^2 k = c^2 p^2 = \frac{cn\pi}{L}.$$

The general solution is

$$G_n(t) = C_n \cos\lambda_n t + D_n \sin\lambda_n t. \tag{5.18}$$

Hence, Equation 5.1 may be written as

$$u_n(x,\ t) = (C_n \cos\lambda_n t + D_n \sin\lambda_n t)\sin\frac{n\pi}{L}x, \quad n = 1,\ 2,\ \ldots \tag{5.19}$$

The entire solution is

$$u(x,\ t) = \sum_{n=1}^{\infty} u_n(x,\ t) = \sum_{n=1}^{\infty} (C_n \cos\lambda_n t + D_n \sin\lambda_n t)\sin\frac{n\pi}{L}x, \quad n = 1,\ 2,\ \ldots \tag{5.20}$$

We have the initial condition

$$u(x,\ 0) = \sum_{n=1}^{\infty} C_n \sin\frac{n\pi}{L}x = f(x) \tag{5.21}$$

Using the Fourier sine series of $f(x)$, we can write C_n as

$$C_n = \frac{2}{L}\int_0^L f(x)\sin\frac{n\pi}{L}x, \quad n = 1, 2, \ldots \tag{5.22}$$

Similarly, using the condition (5.4), we obtain

$$\frac{\partial u}{\partial t}\bigg|_{t=0} = \left[\sum_{n=1}^{\infty}(-C_n\lambda_n \sin\lambda_n t + D_n\lambda_n \cos\lambda_n t)\sin\frac{n\pi}{L}x\right]\bigg|_{t=0}$$

$$= \sum_{n=1}^{\infty} D_n\lambda_n \sin\frac{n\pi}{L}x = g(x). \tag{5.23}$$

Using the Fourier series, we get

$$D_n = \frac{2}{\lambda_n L} \int_0^L g(x) \sin \frac{n\pi}{L} x = \frac{2}{cn\pi} \int_0^L g(x) \sin \frac{n\pi}{L} x =, \quad n = 1, 2, \ldots \tag{5.24}$$

5.2 Numerical Methods

Sometimes, the presence of operating conditions, domain of the problem, coefficients and constants makes the physical problem complicated to investigate. In that case, it is very difficult to analyze and solve the problem by using analytical methods. As such, numerical methods are to be used to investigate such problems. Accordingly, we have presented here two well-known numerical methods, viz. the FDM and the FEM.

5.2.1 Finite Difference Method

Here, finite differences are used for the differentials of the dependent variables appearing in partial differential equations. As such, using some algorithm and standard arithmetic, a digital computer can be employed to obtain a solution. Two methods, viz. the Taylor series expansion and the polynomial representation, are considered in this chapter for approximating the differentials of a function f.

5.2.1.1 Taylor Series Expansion

Given an analytical function $f(x)$, $f(x + \Delta x)$ can be expanded in a Taylor series about x as follows (Hoffmann and Chiang 2000):

$$f(x + \Delta x) = f(x) + (\Delta x) \frac{\partial f}{\partial x} + \frac{(\Delta x)^2}{2!} \frac{\partial^2 f}{\partial x^2} + \frac{(\Delta x)^3}{3!} \frac{\partial^3 f}{\partial x^3} + \cdots$$

$$= \sum_{n=0}^{\infty} \frac{(\Delta x)^n}{n!} \frac{\partial^n f}{\partial x^n} \tag{5.25}$$

The difference approximation $\partial f / \partial x$ can be obtained in the following way:

$$\frac{\partial f}{\partial x} = \frac{f(x + \Delta x) - f(x)}{\Delta x} - \frac{\Delta x}{2!} \frac{\partial^2 f}{\partial x^2} - \frac{(\Delta x)^2}{3!} \frac{\partial^3 f}{\partial x^3} - \cdots \tag{5.26}$$

Considering $O(\Delta x) = -\frac{\Delta x}{2!} \frac{\partial^2 f}{\partial x^2} - \frac{(\Delta x)^2}{3!} \frac{\partial^3 f}{\partial x^3} - \cdots$ in Equation 5.26, we have

$$\frac{\partial f}{\partial x} = \frac{f(x + \Delta x) - f(x)}{\Delta x} + O(\Delta x) \tag{5.27}$$

which is a difference approximation for the first partial derivative of f with respect to x.

If the subscript index i is used to represent the discrete points in the x-direction, Equation 5.27 is written as

$$\left.\frac{\partial f}{\partial x}\right|_i = \frac{f_{i+1} - f_i}{\Delta x} + O(\Delta x). \tag{5.28}$$

Equation 5.28 is known as the first forward difference approximation of $\partial f / \partial x$, which is of order (Δx). So, it is obvious that when the step size decreases, the error term is reduced and hence the accuracy of the approximation increases. Let us consider the Taylor series expansion of $f(x - \Delta x)$ about x.

$$f(x - \Delta x) = f(x) - (\Delta x)\frac{\partial f}{\partial x} + \frac{(\Delta x)^2}{2!}\frac{\partial^2 f}{\partial x^2} - \frac{(\Delta x)^3}{3!}\frac{\partial^3 f}{\partial x^3} + \cdots$$

$$= \sum_{n=0}^{\infty}\left[(-1)^n \frac{(\Delta x)^n}{n!}\right]\frac{\partial^n f}{\partial x^n} \tag{5.29}$$

Solving for $\partial f / \partial x$, we obtain

$$\frac{\partial f}{\partial x} = \frac{f(x) - f(x - \Delta x)}{\Delta x} + O(\Delta x) \tag{5.30}$$

If the subscript index i is used to represent the discrete points in the x-direction, Equation 5.30 is written as

$$\left.\frac{\partial f}{\partial x}\right|_i = \frac{f_i - f_{i-1}}{\Delta x} + O(\Delta x) \tag{5.31}$$

Equation 5.31 is named as the first backward difference approximation of $\partial f / \partial x$ having the order of Δx. Now, consider the Taylor series expansions (5.25) and (5.29), which are repeated here and can be written as

$$f(x + \Delta x) = f(x) + (\Delta x)\frac{\partial f}{\partial x} + \frac{(\Delta x)^2}{2!}\frac{\partial^2 f}{\partial x^2} + \frac{(\Delta x)^3}{3!}\frac{\partial^3 f}{\partial x^3} + \cdots$$

$$f(x - \Delta x) = f(x) - (\Delta x)\frac{\partial f}{\partial x} + \frac{(\Delta x)^2}{2!}\frac{\partial^2 f}{\partial x^2} - \frac{(\Delta x)^3}{3!}\frac{\partial^3 f}{\partial x^3} + \cdots$$

Subtracting Equation 5.29 from Equation 5.25, one obtains

$$f(x + \Delta x) - f(x - \Delta x) = 2(\Delta x)\frac{\partial f}{\partial x} + 2\frac{(\Delta x)^3}{3!}\frac{\partial^3 f}{\partial x^3} + \cdots \tag{5.32}$$

solving for $\partial f / \partial x$,

$$\frac{\partial f}{\partial x} = \frac{f(x+\Delta x) - f(x-\Delta x)}{2\Delta x} + O(\Delta x)^2. \tag{5.33}$$

For the index i, the difference approximation of f with respect to x is

$$\frac{\partial f}{\partial x}\bigg|_i = \frac{f_{i+1} - f_{i-1}}{2\Delta x} + O(\Delta x)^2. \tag{5.34}$$

The representation Equation 5.34 of $\partial f / \partial x$ is known as the central difference approximation of the order $(\Delta x)^2$. Three approximations for the first derivative of $\partial f / \partial x$ have been introduced. Furthermore, the derivations of the finite approximate expressions for the higher-order derivatives are considered.

Again, consider the Taylor series expansion

$$f(x+\Delta x) = f(x) + (\Delta x)\frac{\partial f}{\partial x} + \frac{(\Delta x)^2}{2!}\frac{\partial^2 f}{\partial x^2} + \frac{(\Delta x)^3}{3!}\frac{\partial^3 f}{\partial x^3} + \cdots$$

Expanding by using a Taylor series of $f(x+2\Delta x)$ about x gives the following expansion:

$$f(x+2\Delta x) = f(x) + (2\Delta x)\frac{\partial f}{\partial x} + \frac{(2\Delta x)^2}{2!}\frac{\partial^2 f}{\partial x^2} + \frac{(2\Delta x)^3}{3!}\frac{\partial^3 f}{\partial x^3} + \cdots \tag{5.35}$$

Multiplying Equation 5.25 by 2 and subtracting it from Equation 5.35, we get

$$-2f(x+\Delta x) + f(x+2\Delta x) = -f(x) + (\Delta x)^2\frac{\partial^2 f}{\partial x^2} + (\Delta x)^3\frac{\partial^3 f}{\partial x^3} + \cdots \tag{5.36}$$

Now solving for $\partial^2 f / \partial x^2$, we get

$$\frac{\partial^2 f}{\partial x^2} = \frac{f(x+2\Delta x) - 2f(x+\Delta x) + f(x)}{(\Delta x)^2} + O(\Delta x) \tag{5.37}$$

This equation can be represented as

$$\frac{\partial^2 f}{\partial x^2}\bigg|_i = \frac{f_{i+2} - 2f_{i+1} + f_i}{(\Delta x)^2} + O(\Delta x). \tag{5.38}$$

Equation 5.39 represents the forward difference approximation for the second derivative of f with respect to x and is of the order Δx. Similarly, the approximation for the second derivative can be obtained by using the Taylor series expansions of $f(x-\Delta x)$ and $f(x-2\Delta x)$. The second derivatives are

$$\frac{\partial^2 f}{\partial x^2}\bigg|_i = \frac{f_i - 2f_{i-1} + f_{i-2}}{(\Delta x)^2} + O(\Delta x) \tag{5.39}$$

$$\frac{\partial^2 f}{\partial x^2}\bigg|_i = \frac{f_{i+1} - 2f_i + f_{i-1}}{(\Delta x)^2} + O(\Delta x)^2 \tag{5.40}$$

These equations are backward and central approximations of $\partial^2 f / \partial x^2$, respectively.

5.2.1.2 Finite Difference by Polynomials

The other way for approximating a derivative is to represent the function through a polynomial. The coefficients of the polynomial are obtained by the substitution of a dependent variable from a series of equally spaced points of the independent variables. As such, the approximate values of the derivatives are calculated from the polynomial.

Consider a second-order polynomial as

$$f(x) = Ax^2 + Bx + C \tag{5.41}$$

where A, B and C are constants and we assume the origin at x_i. Thus, $x_i = 0$, $x_{i+1} = \Delta x$ and $x_{i+2} = 2\Delta x$ and so on. The values of function f for corresponding x_i are $f(x_i) = f_i$, $f(x_{i+1}) = f_{i+1}$ and $f(x_{i+2}) = f_{i+2}$.

Thus,

$$f_i = Ax_i^2 + Bx_i + C = C \tag{5.42}$$

$$f_{i+1} = A(x_{i+1})^2 + Bx_{i+1} + C = A(\Delta x)^2 + B(\Delta x) + C \tag{5.43}$$

and

$$f_{i+2} = A(x_{i+2})^2 + Bx_{i+2} + C = A(2\Delta x)^2 + B(2\Delta x) + C. \tag{5.44}$$

From Equations 5.42 through 5.44, it follows that

$$C = f_i$$

$$B = \frac{-f_{i+2} + 4f_{i+1} - 3f_i}{2(\Delta x)}$$

and

$$A = \frac{f_{i+2} - 2f_{i+1} + f_i}{2(\Delta x)^2}.$$

To compute the first derivative of f, one has the following:

$$\frac{\partial f}{\partial x} = 2Ax + B \tag{5.45}$$

At $x_i = 0$, we have

$$\left.\frac{\partial f}{\partial x}\right|_i = B$$

Hence, Equation 5.45 will be

$$\frac{\partial f}{\partial x} = \frac{-f_{i+2} + 4f_{i+1} - 3f_i}{2\Delta x}, \tag{5.46}$$

which is same as the second-order accurate forward difference expression obtained by using Taylor series expansion. Here, one may note that this approximation is classified as second-order accurate for $\partial f/\partial x$, since $\partial^3 f/\partial x^3$ vanishes just as in the accuracy analysis of the Taylor series expansion. The second derivative of f may be calculated as

$$\frac{\partial^2 f}{\partial x^2} = 2A.$$

In a similar way, one may write the following:

$$\frac{\partial^2 f}{\partial x^2} = \frac{f_{i+2} - 2f_{i+1} + f_i}{(\Delta x)^2}$$

and is consistent with the first-order finite difference approximation given by Equation 5.38. Considering the spacing of points i, $i+1$ and $i+2$ as non-identical, a general form of the finite difference approximation of the derivative procedure at $x_i = 0$, $x_{i+1} = \Delta x$ and $x_{i+2} = (1+\alpha)\Delta x$ is as follows.

Here,

$$f_i = C$$

$$f_{i+1} = A(\Delta x)^2 + B(\Delta x) + C$$

and

$$f_{i+2} = A(1+\alpha)^2(\Delta x)^2 + B(1+\alpha)(\Delta x) + C.$$

Consequently,

$$C = f_i$$

$$B = \frac{-f_{i+2} + (1+\alpha)^2 f_{i+1} - (\alpha^2 + 2\alpha) f_i}{\alpha(1+\alpha)\Delta x}$$

and

$$A = \frac{f_{i+2} - (1+\alpha) f_{i+1} + \alpha f_i}{\alpha(1+\alpha)(\Delta x)^2}.$$

Therefore,

$$\frac{\partial f}{\partial x}\bigg|_i = \frac{-f_{i+2} + (1+\alpha)^2 f_{i+1} - \alpha(\alpha+2) f_i}{\alpha(1+\alpha)(\Delta x)},$$

which is a second-order approximation. The second derivative of f is obtained as

$$\frac{\partial^2 f}{\partial x^2} = 2A.$$

As a result,

$$\frac{\partial^2 f}{\partial x^2} = 2\left[\frac{f_{i+2} - (1+\alpha) f_{i+1} + \alpha f_i}{\alpha(1+\alpha)(\Delta x)^2}\right],$$

which is a first-order expression. In the same manner, similar relations for backward and central difference approximations may also be obtained.

5.2.1.3 Finite Difference Equations

Here, the same finite difference approximations are used to replace the derivatives that appear in the partial differential equations (PDEs). Consider an example of PDE involving time (t) and two spatial coordinates x and y. The dependent variable f is $f = f(t, x, y)$.

Assume the governing PDE of the form (Bhat and Chakraverty 2003)

$$\frac{\partial f}{\partial t} = \alpha\left(\frac{\partial^2 f}{\partial x^2} + \frac{\partial^2 f}{\partial y^2}\right) \tag{5.47}$$

where α is any arbitrary constant and is used for illustration purposes.

It is required to approximate the PDE by using the finite difference equation in a domain with equal grid spacing. The Cartesian coordinates x and y are indicated by the subscript indices i and j, and the superscript index n is used to represent time. A first-order finite difference approximation in time and central differencing of second-order accuracy in space is used to represent the time and space differentials, respectively. The spatial grid spacing are Δx and Δy, whereas Δt represents the time step.

It can be noted that the value of f at time level n is known, and the value of f at time level $n+1$ is to be evaluated. So, Equation 5.47 may be expressed at time level n or at time level $n+1$. As a result, two types of formulation are possible. First, let us consider Equation 5.47 at time level n. In this case, a forward difference approximation is used. Hence, we get

$$\frac{\partial f}{\partial t} = \frac{f_{i,j}^{n+1} - f_{i,j}^n}{\Delta t} + O(\Delta t) \tag{5.48}$$

Equation 5.47 for spatial coordinates becomes

$$\frac{\partial^2 f}{\partial x^2} = \frac{f_{i+1,j}^n - 2f_{i,j}^n + f_{i-1,j}^n}{(\Delta x)^2} + O(\Delta x)^2 \tag{5.49}$$

and

$$\frac{\partial^2 f}{\partial y^2} = \frac{f_{i,\,j+1}^n - 2f_{i,\,j}^n + f_{i,\,j-1}^n}{(\Delta y)^2} + O(\Delta y)^2. \tag{5.50}$$

Therefore, the partial differential equation (5.47) in terms of the finite difference formulation can be represented as

$$\frac{f_{i,\,j}^{n+1} - f_{i,\,j}^n}{\Delta t} = \alpha \left[\frac{f_{i+1,\,j}^n - 2f_{i,\,j}^n + f_{i-1,\,j}^n}{(\Delta x)^2} + \frac{f_{i,\,j+1}^n - 2f_{i,\,j}^n + f_{i,\,j-1}^n}{(\Delta y)^2} \right] + O\left[\Delta t, (\Delta x)^2, (\Delta y)^2 \right] \tag{5.51}$$

One may notice that in this formulation, the spatial approximations are applied at time level n.

For the second case, Equation 5.47 is evaluated at time level $n+1$. Therefore, the spatial approximations are at time level $n+1$ and a first-order backward difference approximation in time is employed. As a result, the finite difference formulation takes the following form (Chakraverty 2008):

$$\frac{f_{i,\,j}^{n+1} - f_{i,\,j}^n}{\Delta t} = \alpha \left[\frac{f_{i+1,\,j}^{n+1} - 2f_{i,\,j}^{n+1} + f_{i-1,\,j}^{n+1}}{(\Delta x)^2} + \frac{f_{i,\,j+1}^{n+1} - 2f_{i,\,j}^{n+1} + f_{i,\,j-1}^{n+1}}{(\Delta y)^2} \right] + O\left[\Delta t, (\Delta x)^2, (\Delta y)^2 \right] \tag{5.52}$$

The resulting finite difference equations (5.51) and (5.52) are classified as explicit and implicit formulations, respectively.

The basic distinction between two finite difference equations is the number of unknowns appearing in each equation. The study reveals that Equation 5.51 involves only one unknown, $f_{i,\,j}^{n+1}$, whereas Equation 5.52 involves five unknowns. So, the solution procedures based on explicit and implicit formulations are different in nature. In the explicit formulation, only one unknown appears and may therefore be solved directly at each grid point, whereas in the implicit formulation, more than one unknown exists and therefore the finite difference equation must be written for all the spatial grid points at $n+1$ time level. This represents the same number of equations as the number of unknowns, and those equations are solved simultaneously. It may be noted that the solution of the implicit equation is more difficult than the explicit formulation. Finally, we may say that the implicit formulations are more stable than the explicit formulations.

5.2.2 Finite Element Method

In this section, the finite element method (FEM) has been discussed where the nodes can be spaced in any desired orientation so that a region of any shape can be accommodated. Particularly, closely spaced nodes are used for approximations of curved boundaries. It is not such a difficult task to assign nodes closer together in subregions, where the function is changing rapidly, so it improves the accuracy. Due to the wide adaptability of this method, it is very popular. The complicated application problems are investigated through various finite element analysis (FEA) software in which it has to define the region, set up the equations for all types of boundary conditions and then get the solution.

The basis of FEA is to break up the regions (or domains) of the physical problem into small subregions that are called elements. For example, triangles or rectangles are used to approximate a 2D (two-dimensional) region as well as the curved sides. Whereas in a 3D (three-dimensional) region, pyramids or bricks elements are used. First, the region (2D or 3D) and its elements (triangular or pyramidal) are defined, then the equations for the system are set up and finally investigate the system of equations. The equations are provided some boundary conditions, which are incorporated and solved.

FEA can be used to investigate all three types (parabolic, elliptic and hyperbolic) of partial-differential equations and other eigenvalue problems. A general treatment of FEA to solve ordinary differential equation has been discussed here to give a clear picture.

5.2.2.1 Finite Elements for Ordinary Differential Equations

Following are the steps involved in the FEA and the corresponding method is called the FEM (Gerald and Wheatley 2003, Bhavikatti 2005):

1. Subdivide the domain (interval) [a, b] into n number of subintervals, which are called elements. The nodes of those subintervals are at x_0, x_1, ..., x_n and joins at $x_1, x_2, ..., x_{n-1}$, where $x_0 = a$ and $x_n = b$ are the end nodes. We assume x_i as the nodes of the interval and number the elements from 1 to n, where element 'i' runs from x_{i-1} to x_i. The x_i need not be evenly spaced.

2. Then we apply the Galarkin method to each element separately to interpolate (subject to the differential equation) between the end nodal values, $u(x_{i-1})$ and $u(x_i)$. Here, the u's are approximations to the $y(x_i)$'s, which are the true solution to the differential equation. The nodal values are taken as c's in our adaptation.

3. Use a low-degree polynomial for $u(x)$. Here, a first-degree polynomial has been used, although quadratics or cubics may also be used. (The development for these higher-degree polynomials is more complicated than the polynomial taken here.)

4. Applying the Galarkin method to element 'i', we have a pair of equations in which the unknowns are the nodal values at the ends of element 'i', which are the c's. Performing this for each element, we obtain equations that involve all the nodal values. The nodal values are then combined to give a set of equations that we can solve for the unknown nodal values. (The process of combining the separate element equations is called assembling the system.)

5. These equations are adjusted for the boundary conditions and solved to get approximations to $y(x)$ at the nodal points. Finally, we obtain intermediate values for $y(x)$ by linear interpolation.

Consider the following differential equation and its finite element development. It involves several steps and each step is straightforward. The differential equation that will be investigated is

$$y'' + Q(x)y = F(x), \tag{5.53}$$

subject to the boundary conditions at $x = a$ and $x = b$.

Step 1

Subdivide the interval [a, b] into n elements, as discussed earlier. Focus on element 'i' that runs from x_{i-1} to x_i. For simplicity, we call the left node L and the right node R.

Step 2

We can write $u(x)$ for element 'i' as

$$u(x) = c_L N_L + c_R N_R = c_L \frac{x-R}{L-R} + c_R \frac{x-L}{R-L}$$

$$= c_L \frac{x-R}{-h_i} + c_R \frac{x-L}{h_i}. \tag{5.54}$$

The N's in Equation 5.54 are first-degree (linear) Lagrangian polynomials. These shape functions are often called hat functions due to the orientation.

The shape functions N's are nothing but the functions of x and the values vary (from unity to zero) from x_L to x_R. It may be noted that the c's in Equation 5.54 are independent of x.

Step 3

Applying the Galarkin method to element 'i', the residual becomes

$$R(x) = y'' + Qy - F = u'' + Qu - F, \tag{5.55}$$

where $u(x)$ is substituted in place of $y(x)$. The Galarkin method sets the integral of R weighted with each of the N's (over the length of the element) to zero.

$$\int_L^R N_L R(x) dx = 0, \tag{5.56}$$

$$\int_L^R N_R R(x) dx = 0. \tag{5.57}$$

Now expanding Equations 5.56 and 5.57, we get

$$\int_L^R u'' N_L dx + \int_L^R Qu N_L dx - \int_L^R F N_L dx = 0. \tag{5.58}$$

$$\int_L^R u'' N_R dx + \int_L^R Qu N_R dx - \int_L^R F N_R dx = 0. \tag{5.59}$$

Step 4

Now, transform Equation 5.58 by applying integration by parts to the first integral. In the second integral, we will take Q out from the integrand as Q_{int}, an average value within the element. Similarly, take F outside the third integral and it will be denoted as F_{int}. Then Equation 5.58 becomes

$$-\int_L^R \left(\frac{du}{dx}\right)\left(\frac{dN_L}{dx}\right)dx + Q_{int}\int_L^R uN_L dx - F_{int}\int_L^R N_L dx + N_L\frac{du}{dx}\bigg|_{x=R} - N_L\frac{du}{dx}\bigg|_{x=L} = 0 \qquad (5.60)$$

In the last two terms of Equation 5.60, $N_L = 1$ at L and is zero at R, hence Equation 5.60 can be simplified as

$$-\int_L^R \left(\frac{du}{dx}\right)\left(\frac{dN_L}{dx}\right)dx + Q_{int}\int_L^R uN_L dx - F_{int}\int_L^R N_L dx - \frac{du}{dx}\bigg|_{x=L} = 0 \qquad (5.61)$$

In a similar fashion, Equation 5.59 gives

$$-\int_L^R \left(\frac{du}{dx}\right)\left(\frac{dN_R}{dx}\right)dx + Q_{int}\int_L^R uN_R dx - F_{int}\int_L^R N_R dx - \frac{du}{dx}\bigg|_{x=R} = 0 \qquad (5.62)$$

Step 5

Using Equations 5.54, 5.61 and 5.62, we get

$$\int_L^R \left(\frac{du}{dx}\right)\left(\frac{dN_L}{dx}\right)dx = \int_L^R \left[\frac{-c_L}{h_i} + \frac{c_R}{h_i}\right]\left(\frac{-1}{h_i}\right)dx = \left(\frac{c_L}{h_i^2} - \frac{c_R}{h_i^2}\right)\int_L^R dx$$

$$= \left(\frac{1}{h_i}\right)c_L - \left(\frac{1}{h_i}\right)c_R \qquad (5.63)$$

Now,

$$-Q_{int}\int_L^R (c_L N_L + c_R N_R)N_L dx = -c_L Q_{int}\int_L^R (N_L)^2 dx - c_R Q_{int}\int_L^R N_R N_L dx$$

$$= -c_L Q_{int}\int_L^R \left(\frac{x-R}{h_i}\right)^2 dx - c_R Q_{int}\int_L^R \left(\frac{x-L}{h_i}\right)\left(\frac{x-R}{-h_i}\right)$$

$$= -\left(Q_{int} * \frac{h_i}{3}\right)c_L - \left(Q_{int} * \frac{h_i}{6}\right)c_R$$

and

$$F_{int} \int_{L}^{R} N_L dx = F_{int} \int_{L}^{R} \frac{x-R}{-h_i} dx = F_{int} * \frac{h_i}{2}$$

In the same manner, we also get

$$\int_{L}^{R} \left(\frac{du}{dx}\right)\left(\frac{dN_R}{dx}\right) dx = -\left(\frac{1}{h_i}\right)c_L + \left(\frac{1}{h_i}\right)c_R \tag{5.64}$$

Now,

$$-Q_{int} \int_{L}^{R} (c_L N_L + c_R N_R) N_L dx = -\left(Q_{int} * \frac{h_i}{6}\right)c_L - \left(Q_{int} * \frac{h_i}{3}\right)c_R \tag{5.65}$$

and

$$F_{int} \int_{L}^{R} N_R dx = F_{int} * \frac{h_i}{2}. \tag{5.66}$$

Step 6

Substitute the values in Step 5 and rearrange to give two linear equations of the unknown c_L and c_R!

$$\left(\frac{1}{h_i} - \frac{Q_{int}h_i}{3}\right)c_L + \left(\frac{-1}{h_i} - \frac{Q_{int}h_i}{6}\right)c_R = \frac{-F_{int}h_i}{2} - \frac{du}{dx}\bigg|_{x=L} \tag{5.67}$$

$$\left(\frac{-1}{h_i} - \frac{Q_{int}h_i}{6}\right)c_L + \left(\frac{1}{h_i} - \frac{Q_{int}h_i}{3}\right)c_R = \frac{-F_{int}h_i}{2} + \frac{du}{dx}\bigg|_{x=R} \tag{5.68}$$

The pair of Equations 5.67 and 5.68 is called element equations. The same procedure is to be adapted for each element to get n such pairs.

Step 7

Furthermore, combine (assemble) all the element equations together to form a system of linear equations for the problem. Point R in element 'i' is certainly the same as point L in element '$i + 1$' and the c's as c_0, c_1, ... , c_n. We also notice that the gradient (du/dx) must be the same on either sides of the join of the elements. As such, $(du/dx)_{x=R}$ for element 'i' equals to $(du/dx)_{x=L}$ in element '$i + 1$'.

Finally, the results for repeating Step 7 in the set of $n + 1$ equations (numbered from 0 to n) give

$$[K]\{c\} = \{b\} \tag{5.69}$$

where the diagonal elements of $[K]$ are

$$\text{diag}\begin{bmatrix} \left(\dfrac{1}{h_i}-Q_{int,1}*\dfrac{h_1}{3}\right) \\[2ex] \left(\dfrac{1}{h_i}-Q_{int,i}*\dfrac{h_i}{3}\right)+\left(\dfrac{1}{h_{i+1}}-Q_{int,i+1}*\dfrac{h_{i+1}}{3}\right) \\[1ex] \vdots \\[1ex] \left(\dfrac{1}{h_i}-Q_{int,i}*\dfrac{h_i}{3}\right)+\left(\dfrac{1}{h_{i+1}}-Q_{int,i+1}*\dfrac{h_{i+1}}{3}\right) \\[2ex] \left(\dfrac{1}{h_n}-Q_{int,n}*\dfrac{h_n}{3}\right) \end{bmatrix}$$

where

$$\left(\frac{1}{h_i}-Q_{int,1}*\frac{h_1}{3}\right) \text{ in row 0}$$

$$\left(\frac{1}{h_i}-Q_{int,i}*\frac{h_i}{3}\right)+\left(\frac{1}{h_{i+1}}-Q_{int,i+1}*\frac{h_{i+1}}{3}\right) \text{ in rows 1 to } n-1$$

$$\left(\frac{1}{h_n}-Q_{int,n}*\frac{h_n}{3}\right) \text{ in row } n$$

and elements above and to the left of the diagonal in rows $1-n$ are

$$\left(\frac{-1}{h_i}-Q_{int,i}*\frac{h_i}{6}\right).$$

The elements of $\{c\}$ are c_i, $i=0$ to n.

The elements of $\{b\}$ are

$$-F_{int,1}*\frac{h_1}{2}-\left(\frac{du}{dx}\right)_{x=a} \text{ in row 0,}$$

$$-F_{int,i}*\frac{h_i}{2}-F_{int,i+1}*\frac{h_{i+1}}{2} \text{ in rows 1 to } n-1$$

$$-F_{int,n}*\frac{h_n}{2}+\left(\frac{du}{dx}\right)_{x=b} \text{ in row } n.$$

In the preceding equations, $Q_{int,i}$ and $F_{int,i}$ are values of Q and F at the midpoints of element 'i'.

Step 8

Finally, adjust the set of equations of Step 6 and incorporate the boundary conditions. There are three cases out of which two cases have been discussed here. Case 1, a Dirichlet condition is specified as $-y(a)$ = constant [and/or $y(b)$ = constant]. Case 2, a Neumann condition is specified as $-dy/dx$ = constant at $x=a$ and/or $x=b$. (If $Q=0$, we cannot have a Neumann condition at both ends, because the solution would be known only within an additive constant.) [We leave Case 3, mixed conditions, which is a modification of Case 2.]

Case 1 Dirichlet Condition

In this case, c is known at the end node. Suppose this is $y(a)=K_a$, then the equation in row 0 is redundant, and so we remove it from the set of equations of Step 6. In the next row, we move $k_{10} * K_a$ to the right-hand side (subtracting this from the element computed in Step 6). If the condition is $y(b)=K_b$, we do the same but with the last and next to last equations.

Case 2 Neumann Condition

In this case, c is not known at the end node. Suppose the condition is $dy/dx=K_a$ at $x=a$, we retain the equation in row 0 and substitute the given value of dy/dx into the right-hand side. If the condition is $dy/dx=K_b$ at $x=b$, we do the same with the last equation.

Step 9

Solve the set of equations for the unknowns c's after adjusting, in Step 8, for the boundary conditions. The c's are approximations to $y(x)$ at the nodes. If the intermediate values of y are needed between the nodes, we obtain them by linear interpolation.

Bibliography

Bhat, R. and Chakraverty, S. 2003. *Numerical Analysis in Engineering*. Alpha Science International Ltd., U.K.

Bhavikatti, S. S. 2005. *Finite Element Analysis*. New Age International, New Delhi, India.

Chakraverty, S. 2008. *Vibration of Plates*. CRC Press, Boca Rato, FL.

Gerald, C. F. and Wheatley, P. O. 2003. *Applied Numerical Analysis*, 7th edn. Pearson, Noida, India.

Hoffmann, K. A. and Chiang, S. T. 2000. *Computational Fluid Dynamics*, 4th edn. Engineering Education System, Wichita, USA.

Kreyszig, E. 2010. *Advanced Engineering Mathematics*. Wiley, USA.

6

Uncertain One-Group Model

The scattering of neutron collision inside a reactor depends on the geometry of the reactor, the diffusion coefficient and the absorption coefficient. In general, these parameters are not crisp (exact) and hence we may get an uncertain neutron diffusion equation. In this chapter, the uncertain neutron diffusion equation for a bare square homogeneous reactor is discussed. The uncertain governing differential equation is modelled by a modified fuzzy finite element method (FFEM). Using the modified FFEM, the obtained eigenvalues and effective multiplication factors are studied. The corresponding results are compared with the classical finite element method (FEM) in special cases and various uncertain results are explained.

6.1 Background

Uncertainty plays a vital role in various fields of engineering and science. These uncertainties occur due to incomplete data, impreciseness, vagueness, experimental error and different operating conditions influenced by the system. Different authors have proposed various methods to handle this uncertainty. They have used the probabilistic or statistical method as a tool to handle uncertain parameters. In this context, the Monte Carlo method is an alternative method, which is based on the statistical simulation of the random numbers generated on the basis of a specific sampling distribution. Monte Carlo methods have been used to solve the neutron diffusion equation with variable parameters. As such, Nagaya et al. (2010) implemented the Monte Carlo method to estimate the effective delayed neutron fraction β_{eff}. Furthermore, Nagaya and Mori (2011) proposed a new method to estimate the effective delayed neutron fraction β_{eff} in Monte Carlo calculations. In that article, the eigenvalue method was jointly used with the differential operator and correlated sampling techniques, whereas Shi and Petrovic (2011) used Monte Carlo methods to solve 1D two-group problems and then they proved its validity for these problems. Sjenitzer and Hoogenboom (2011) proposed an analytical procedure to compute the variance of the neutron flux in a simple model of a fixed-source calculation. Recently, Yamamoto (2012) investigated the neutron leakage effect specified by buckling to generate group constants for use in reactor core designs using the Monte Carlo method.

As such, in this process, we need a good number of observed data or experimental results to analyze the problem. Sometimes, it may not be possible to get a large number of data. As regards this process, Zadeh (1965) proposed an alternate idea, viz. the fuzzy approach, to handle uncertain and imprecise variables. Accordingly, we may use interval or fuzzy parameters to take care of the uncertainty. In general, traditional interval/fuzzy arithmetic is complicated to investigate the problem. In this context, a new technique for fuzzy arithmetic is developed to overcome such difficulty, which is proposed by Chakraverty and Nayak (2012). A few authors have investigated the mentioned problem. In this respect, Biswas et al. (1976) have given a method of generating stiffness matrices for the solution of the multigroup diffusion equation by a natural coordinate system. Azekura (1980) has also proposed a new representation of the finite element solution technique for neutron diffusion equations. The author has applied the technique to two types of one-group neutron diffusion equations to test its accuracy. Furthermore, Cavdar and Ozgener (2004) developed a finite element-boundary element-hybrid method for one- or two-group neutron diffusion calculations. In their article, a linear or bilinear finite element formulation for the reactor core and a linear boundary element technique for the reflector, which are combined through interface continuity conditions, constitute the basis of the developed method. Dababneh et al. (2011) formulated an alternative analytical solution of the neutron diffusion equation for both infinite and finite cylinders of fissile material using the homotopy perturbation method, whereas Rokrok et al. (2012) applied the element-free Galerkin (EFG) method to solve the neutron diffusion equation in X–Y geometry. From this, it is clear that neutron diffusion equations are solved using the FEM in the presence of crisp parameters only.

But the presence of uncertain parameters makes the system uncertain, and we get uncertain governing differential equations. In this context, uncertain fuzzy parameters are considered to solve heat conduction problems using the FEM, and we call it the FFEM. Bart et al. (2011) solved the uncertain solution of the heat conduction problem. In this article, authors made a good comparison between the response surface method and other methods. Recently, Chakraverty and Nayak (2012) also solved the interval/fuzzy distribution of temperature along a cylindrical rod. Here, they have presented a modified form of FFEM. The involved fuzzy numbers are changed into intervals through α-cut. Then, the intervals are transformed into crisp form by using some transformations. Crisp representations of intervals are defined by a symbolic parameterization. Traditional interval arithmetic is modified using the crisp representation of intervals. The interval arithmetic is extended for fuzzy numbers, and the developed fuzzy arithmetic is used as a tool for the uncertain FFEM. Consequently, this method is used to solve the one-group neutron diffusion equation, and the critical eigenvalues and effective multiplication factors are studied in detail. Hence, it may be used as a tool to solve different types of neutron diffusion problems for various types of nuclear reactors.

It is already mentioned that the uncertainty is considered here in terms of fuzzy or interval. In order to handle the uncertainty while using the FEM, we must formulate the FEM in uncertainty term (fuzzy or interval) using the fuzzy or interval computation. We have first formulated the following problem as traditional FEM for the sake of completeness. Then, the problem has been handled considering the uncertainty, and the example problem has been taken as a square homogeneous bare reactor.

6.2 Formulation of the Problem

As it is known that the principle of neutron conservation can be expressed in a simple form for a system of mono-energetic neutrons, one-group equations can be analyzed by considering the series of one-group equations.

The standard functional for the one-group diffusion equation may be written as

$$I(\phi) = \frac{1}{2} \iint\limits_{R} \left[D\left(\frac{\partial \phi}{\partial x}\right)^2 + D\left(\frac{\partial \phi}{\partial y}\right)^2 + \sigma\phi^2 - 2S\phi \right] dxdy \tag{6.1}$$

where
  ϕ is the neutron flux
  D is the diffusion coefficient
  σ is the absorption coefficient
  S is the source term

In the traditional FEM, the domain of the problem is divided into a number of subdomains and each is called an element. For each element, we may find the functional, and similarly, for the entire domain, the functional can be found by summing each functional element wise. This procedure may be written in the following way.

First, the domain R may be represented as

$$R = \sum_{e=1}^{n} R^e \tag{6.2}$$

and the functional $I(\phi)$ is defined as

$$I(\phi) = \sum_{e=1}^{n} I^e(\phi) \tag{6.3}$$

where
  n is the total number of elements
  $I^e(\phi)$ denotes the contribution of the element e to the functional $I(\phi)$

Now, Equation 6.1 for each elemental functional may be written as

$$I^e(\phi) = \frac{1}{2} \iint\limits_{R} \left[D^e\left(\frac{\partial \phi^e}{\partial x}\right)^2 + D^e\left(\frac{\partial \phi^e}{\partial y}\right)^2 + \sigma^e\phi^2 - 2S^e\phi^e \right] dxdy \tag{6.4}$$

For each element e, the scalar flux ϕ^e is approximated by a piece-wise interpolation polynomial. Depending on the interpolation polynomial, stiffness matrices are obtained by minimizing the elemental functional $I^e(\phi)$. The stiffness matrices are assembled, and finally, we get the algebraic form, which is represented as

$$[K]\{\phi\} = \{Q\} \tag{6.5}$$

where
  $[K]$ is the assembled stiffness matrix corresponding to the leakage and absorption terms
  $\{Q\}$ is the assembled force vector for the source term

In general, when neutrons undergo scattering, the neutron transport equation involves uncertainty. The uncertainty occurs due to the imprecise value of the operating parameters, viz. geometry, diffusion and absorption coefficients. Considering uncertain parameters as fuzzy, we now investigate the uncertain spectrum of the neutron flux distribution. Accordingly, we now formulate the FFEM with the linear triangular fuzzy element discretization of the domain.

Let us write the coordinates of linear triangular elements in fuzzy form as

$$
\begin{aligned}
\tilde{x} &= L_1\tilde{x}_1 + L_2\tilde{x}_2 + L_3\tilde{x}_3; \\
\tilde{y} &= L_1\tilde{y}_1 + L_2\tilde{y}_2 + L_3\tilde{y}_3; \\
\tilde{L} &= \tilde{L}_1 + \tilde{L}_2 + \tilde{L}_3
\end{aligned}
\tag{6.6}
$$

where $\tilde{L}_i$ $(i = 1,\ 2,\ 3)$ are non-dimensionalized coordinates.

Equation 6.6 may be written in matrix form

$$
\begin{bmatrix} 1 & 1 & 1 \\ \tilde{x}_1 & \tilde{x}_2 & \tilde{x}_3 \\ \tilde{y}_1 & \tilde{y}_2 & \tilde{y}_3 \end{bmatrix} \begin{Bmatrix} \tilde{L}_1 \\ \tilde{L}_2 \\ \tilde{L}_3 \end{Bmatrix} = \begin{Bmatrix} 1 \\ \tilde{x} \\ \tilde{y} \end{Bmatrix}
$$

$$
\Rightarrow \begin{Bmatrix} \tilde{L}_1 \\ \tilde{L}_2 \\ \tilde{L}_3 \end{Bmatrix} = \begin{bmatrix} 1 & 1 & 1 \\ \tilde{x}_1 & \tilde{x}_2 & \tilde{x}_3 \\ \tilde{y}_1 & \tilde{y}_2 & \tilde{y}_3 \end{bmatrix}^{-1} \begin{Bmatrix} 1 \\ \tilde{x} \\ \tilde{y} \end{Bmatrix}
$$

$$
\Rightarrow \begin{Bmatrix} \tilde{L}_1 \\ \tilde{L}_2 \\ \tilde{L}_3 \end{Bmatrix} = \frac{1}{2\tilde{\Delta}} \begin{bmatrix} \tilde{x}_2\tilde{y}_3 - \tilde{x}_3\tilde{y}_2 & \tilde{y}_2 - \tilde{y}_3 & \tilde{x}_3 - \tilde{x}_2 \\ \tilde{x}_3\tilde{y}_1 - \tilde{x}_1\tilde{y}_3 & \tilde{y}_3 - \tilde{y}_1 & \tilde{x}_1 - \tilde{x}_3 \\ \tilde{x}_1\tilde{y}_2 - \tilde{x}_2\tilde{y}_1 & \tilde{y}_1 - \tilde{y}_2 & \tilde{x}_2 - \tilde{x}_1 \end{bmatrix} \begin{Bmatrix} 1 \\ \tilde{x} \\ \tilde{y} \end{Bmatrix}
$$

where the area of the fuzzy triangle is $\tilde{\Delta} = \dfrac{1}{2} \begin{vmatrix} 1 & 1 & 1 \\ \tilde{x}_1 & \tilde{x}_2 & \tilde{x}_3 \\ \tilde{y}_1 & \tilde{y}_2 & \tilde{y}_3 \end{vmatrix}$.

We now denote

$$
\begin{aligned}
\tilde{a}_1 &= \tilde{x}_3 - \tilde{x}_2, \quad \tilde{a}_2 = \tilde{x}_1 - \tilde{x}_3, \quad \tilde{a}_3 = \tilde{x}_2 - \tilde{x}_1; \\
\tilde{b}_1 &= \tilde{y}_2 - \tilde{y}_3, \quad \tilde{b}_2 = \tilde{y}_3 - \tilde{y}_1, \quad \tilde{b}_3 = \tilde{y}_1 - \tilde{y}_2; \\
\tilde{c}_1 &= \tilde{x}_2\tilde{y}_3 - \tilde{x}_3\tilde{y}_2, \quad \tilde{c}_2 = \tilde{x}_3\tilde{y}_1 - \tilde{x}_1\tilde{y}_3, \quad \tilde{c}_3 = \tilde{x}_1\tilde{y}_2 - \tilde{x}_2\tilde{y}_1.
\end{aligned}
$$

If $\tilde{\phi}$ is the flux distribution, then it may be written as

$$
\tilde{\phi} = \tilde{L}_1\tilde{\phi}_1 + \tilde{L}_2\tilde{\phi}_2 + \tilde{L}_3\tilde{\phi}_3.
\tag{6.7}
$$

The differentiation and integration formulae are then given by

$$\frac{\partial}{\partial \tilde{x}} = \sum_{i=1}^{3} \frac{\tilde{b}_i}{2\tilde{\Delta}} \frac{\partial}{\partial \tilde{L}_i}, \quad \frac{\partial}{\partial \tilde{y}} = \sum_{i=1}^{3} \frac{\tilde{a}_i}{2\tilde{\Delta}} \frac{\partial}{\partial \tilde{L}_i} \quad \text{and} \quad \iint_R \tilde{L}_1^p \tilde{L}_2^q \tilde{L}_3^r \, d\tilde{\Delta} = \frac{p!q!r!}{(p+q+r+2)!}(2\tilde{\Delta}).$$

Hence,

$$\frac{\partial \tilde{\phi}^{(e)}}{\partial \tilde{x}} = \frac{1}{2\tilde{\Delta}}\left\{\tilde{b}_1\tilde{\phi}_1 + \tilde{b}_2\tilde{\phi}_2 + \tilde{b}_3\tilde{\phi}_3\right\} = \begin{bmatrix} \dfrac{\tilde{b}_1}{2\tilde{\Delta}} & \dfrac{\tilde{b}_2}{2\tilde{\Delta}} & \dfrac{\tilde{b}_3}{2\tilde{\Delta}} \end{bmatrix}\begin{Bmatrix} \tilde{\phi}_1 \\ \tilde{\phi}_2 \\ \tilde{\phi}_3 \end{Bmatrix}$$

Similarly,

$$\frac{\partial \tilde{\phi}^{(e)}}{\partial \tilde{y}} = \frac{1}{2\tilde{\Delta}}\left\{\tilde{a}_1\tilde{\phi}_1 + \tilde{a}_2\tilde{\phi}_2 + \tilde{a}_3\tilde{\phi}_3\right\} = \begin{bmatrix} \dfrac{\tilde{a}_1}{2\tilde{\Delta}} & \dfrac{\tilde{a}_2}{2\tilde{\Delta}} & \dfrac{\tilde{a}_3}{2\tilde{\Delta}} \end{bmatrix}\begin{Bmatrix} \tilde{\phi}_1 \\ \tilde{\phi}_2 \\ \tilde{\phi}_3 \end{Bmatrix}.$$

Using this formulation, one may get the leakage and absorption stiffness matrices. Accordingly, corresponding stiffness matrices of each element for the leakage and absorption terms are found to be

$$\left[\tilde{K}_1\right] = \frac{\tilde{D}^{(e)}}{4\tilde{\Delta}}\begin{bmatrix} \tilde{a}_1^2 + \tilde{b}_1^2 & \tilde{a}_1\tilde{a}_2 + \tilde{b}_1\tilde{b}_2 & \tilde{a}_1\tilde{a}_3 + \tilde{b}_1\tilde{b}_3 \\ \tilde{a}_1\tilde{a}_2 + \tilde{b}_1\tilde{b}_2 & \tilde{a}_2^2 + \tilde{b}_2^2 & \tilde{a}_2\tilde{a}_3 + \tilde{b}_2\tilde{b}_3 \\ \tilde{a}_1\tilde{a}_3 + \tilde{b}_1\tilde{b}_3 & \tilde{a}_2\tilde{a}_3 + \tilde{b}_2\tilde{b}_3 & \tilde{a}_3^2 + \tilde{b}_3^2 \end{bmatrix} \quad \text{and} \quad \left[\tilde{K}_2\right] = \frac{\tilde{\sigma}^{(e)}\tilde{\Delta}}{12}\begin{bmatrix} 2 & 1 & 1 \\ 1 & 2 & 1 \\ 1 & 1 & 2 \end{bmatrix}, \text{respectively.}$$

The source vector $\{\tilde{f}\}$ for each element may be written as

$$\{\tilde{f}\} = \frac{\tilde{S}^{(e)}\tilde{\Delta}}{3}\begin{Bmatrix} 1 \\ 1 \\ 1 \end{Bmatrix}.$$

The limit method for fuzzy arithmetic in terms of α-cut has been used here for FEM. As such, using FFEM (Nayak and Chakraverty 2013), the uncertain fuzzy parameters are handled. The schematic diagram is presented in Figure 4.1, which gives an overall idea to encrypt the process of modified FFEM. It involves three steps such as the input, the output and the hidden layer. The uncertain (fuzzy) parameters are accumulated through a process, viz. FFEM, which plays the role of the hidden layer. Using the fuzzy input parameters, the hidden layer performs the fuzzy finite element procedure and gives fuzzy solutions in terms of fuzzy as output. In Figure 4.1, triangular fuzzy numbers have been considered as input parameters. The alpha (α)-level representation of two fuzzy sets $\tilde{X}$ and $\tilde{Y}$ with their triangular membership functions are operated through fuzzy arithmetic operation

(Chakraverty and Nayak 2012). The deterministic value is obtained for α_4 level of fuzzy sets, whereas for α_1, α_2 and α_3 levels we get interval values. The output can be generated by taking the union of all possible combinations of the alpha level.

6.3 Case Study 1

The governing differential equation in crisp form for the bare homogeneous reactor (Glasstone and Sesonke 2004) is as follows:

$$D\nabla^2\phi + S = \Sigma_a\phi \tag{6.8}$$

The boundary condition is $\phi(x, \pm1.5) = 0 = \phi(\pm1.5, y)$, and it is solved first by the classical (traditional) FEM for the sake of completeness, and then the FFEM is presented. Here, the square homogeneous region has been divided into 18 and 72 elements as shown in Figures 6.1a and b and 6.2a and b. Two different types of discretizations are considered and the results are compared.

Initially, the eigenvalues and corresponding effective multiplication factors are investigated when the involved parameters, viz. diffusion coefficient (D) and absorption coefficient (σ), are crisp. The different values of these parameters are given in Table 6.1. The observed results for two different types of discretizations are presented in Tables 6.2 through 6.7.

When neutrons undergo diffusion, they suffer scattering collisions with the nuclei assumed to be initially stationary, and as a result, a typical neutron trajectory consists of a number of short-path elements. These are scattering free paths which are uncertain in nature. As a result, the diffusion coefficient will lie in an uncertain region and may become fuzzy. Similarly, the absorption coefficient may also be taken as fuzzy. Here, we have taken two different types of fuzzy numbers (TFN and TRFN) to handle these uncertainties. The uncertain values along with the crisp values are given in Table 6.1.

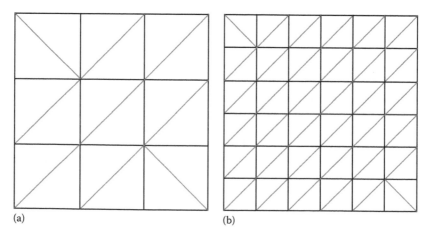

(a)   (b)

FIGURE 6.1
Domain is discretized (type 1) into (a) 18 elements and (b) 72 elements.

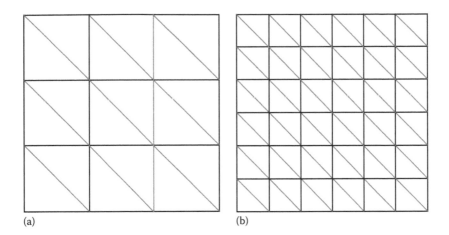

FIGURE 6.2
Domain is discretized (type 2) into (a) 18 elements and (b) 72 elements.

TABLE 6.1

Crisp and Fuzzy Values of the Involved Parameters

Parameters	Crisp Value	TFN	TRFN
Diffusion coefficient	1	$[0.5 + 0.5\alpha, 1.5 - 0.5\alpha]$	$[0.5 + 0.3\alpha, 1.5 - 0.3\alpha]$
Absorption coefficient	1	$[0.5 + 0.5\alpha, 1.5 - 0.5\alpha]$	$[0.5 + 0.3\alpha, 1.5 - 0.3\alpha]$

TABLE 6.2

Comparison of Eigenvalues When $\sigma = 1$ and $D = [0.5 + 0.5\alpha, 1.5 - 0.5\alpha]$

Number of Elements	Classical FEM (Figure 6.1a and b)	Proposed Fuzzy FEM (Figure 6.1a and b)	Classical FEM (Figure 6.2a and b)	Proposed Fuzzy FEM (Figure 6.2a and b)
18	2.6667	$[1.3333 + 1.3334\alpha, 4 - 1.3333\alpha]$	2.8195	$[1.4097 + 1.4098\alpha, 4.2293 - 1.4098\alpha]$
72	2.3426	$[1.1713 + 1.1713\alpha, 3.5140 - 1.1714\alpha]$	2.3454	$[1.1727 + 1.1727\alpha, 3.5181 - 1.1727\alpha]$

TABLE 6.3

Comparison of Eigenvalues When $D = 1$ and $\sigma = [0.5 + 0.5\alpha, 1.5 - 0.5\alpha]$

Number of Elements	Classical FEM (Figure 6.1a and b)	Proposed Fuzzy FEM (Figure 6.1a and b)	Classical FEM (Figure 6.2a and b)	Proposed Fuzzy FEM (Figure 6.2a and b)
18	2.6667	$[1.7778 + 0.48887\alpha, 5.3333 - 2.6666\alpha]$	2.8195	$[1.8797 + 0.9398\alpha, 5.6391 - 2.8196\alpha]$
72	2.3426	$[1.5618 + 0.7808\alpha, 4.6853 - 2.3427\alpha]$	2.3454	$[1.5636 + 0.7818\alpha, 4.6908 - 2.3454\alpha]$

TABLE 6.4

Comparison of Eigenvalues When $D = [0.5 + 0.5\alpha, 1.5 - 0.5\alpha]$ and $\sigma = [0.5 + 0.5\alpha, 1.5 - 0.5\alpha]$

Number of Elements	Classical FEM (Figure 6.1a and b)	Proposed Fuzzy FEM (Figure 6.1a and b)	Classical FEM (Figure 6.2a and b)	Proposed Fuzzy FEM (Figure 6.2a and b)
18	2.6667	$[0.889 + 1.7777\alpha,$ $8 - 5.3333\alpha]$	2.8195	$[0.9398 + 1.8797\alpha,$ $8.4587 - 5.6392\alpha]$
72	2.3426	$[0.7809 + 1.5617\alpha,$ $7.0279 - 4.6853\alpha]$	2.3454	$[0.7818 + 1.5636\alpha,$ $7.0363 - 4.6909\alpha]$

TABLE 6.5

Comparison of Eigenvalues When $\sigma = 1$ and $D = [0.5 + 0.3\alpha, 1.5 - 0.3\alpha]$

Number of Elements	Classical FEM (Figure 6.1a and b)	Proposed Fuzzy FEM (Figure 6.1a and b)	Classical FEM (Figure 6.2a and b)	Proposed Fuzzy FEM (Figure 6.2a and b)
18	2.6667	$[1.3333 + 0.8\alpha,$ $4 - 0.8\alpha]$	2.8195	$[1.4097 + 0.8459\alpha,$ $4.2293 - 0.8458\alpha]$
72	2.3426	$[1.1713 + 0.7028\alpha,$ $3.5140 - 0.7028\alpha]$	2.3454	$[1.1727 + 0.7036\alpha,$ $3.5181 - 0.7036\alpha]$

TABLE 6.6

Comparison of Eigenvalues When $D = 1$ and $\sigma = [0.5 + 0.3\alpha, 1.5 - 0.3\alpha]$

Number of Elements	Classical FEM (Figure 6.1a and b)	Proposed Fuzzy FEM (Figure 6.1a and b)	Classical FEM (Figure 6.2a and b)	Proposed Fuzzy FEM (Figure 6.2a and b)
18	2.6667	$[1.7778 + 0.4444\alpha,$ $5.3333 - 2\alpha]$	2.8195	$[1.8797 + 0.4699\alpha,$ $5.6391 - 2.1147\alpha]$
72	2.3426	$[1.5618 + 0.3904\alpha,$ $4.6853 - 1.757\alpha]$	2.3454	$[1.5636 + 0.3909\alpha,$ $4.6908 - 1.759\alpha]$

TABLE 6.7

Comparison of Eigenvalues When $D = [0.5 + 0.3\alpha, 1.5 - 0.3\alpha]$ and $\sigma = [0.5 + 0.3\alpha, 1.5 - 0.3\alpha]$

Number of Elements	Classical FEM (Figure 6.1a and b)	Proposed Fuzzy FEM (Figure 6.1a and b)	Classical FEM (Figure 6.2a and b)	Proposed Fuzzy FEM (Figure 6.2a and b)
18	2.6667	$[0.889 + 0.8888\alpha,$ $8 - 4\alpha]$	2.8195	$[0.9398 + 0.9399\alpha,$ $8.4587 - 4.2293\alpha]$
72	2.3426	$[0.7809 + 0.7809\alpha,$ $7.0279 - 3.5139\alpha]$	2.3454	$[0.7818 + 0.7818\alpha,$ $7.0363 - 3.5181\alpha]$

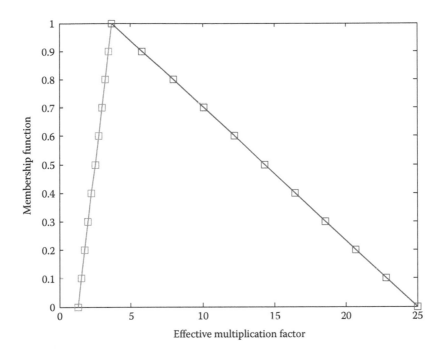

FIGURE 6.3
TFN for Figure 6.1a.

The uncertain values are used in the FFEM and eigenvalues are obtained. The uncertain fuzzy eigenvalues under different considerations are shown in Tables 6.2 through 6.7.

In view of these tabulated eigenvalues, the corresponding effective multiplication factors ($\tilde{k}_{eff}$) are plotted. These are given pictorially in Figures 6.3 through 6.10.

6.3.1 Discussion

To study the eigenvalue problem for the corresponding one-group neutron diffusion equation, we have considered two different types of discretizations of a bare square homogeneous reactor. Initially, the eigenvalue problem is solved by the classical FEM for crisp parameters and the following observations are reported:

- We find that there is a variation in eigenvalues when different discretizations with the same number of elements are taken. Here, it is observed that a better approximation to eigenvalues are found for Figure 6.1a and b.
- In Figure 6.1a and b, there are two distinct nodes where the contribution values towards the stiffness matrix are more as compared to Figure 6.2a and b.
- It is observed that if the contribution value towards the stiffness matrix increases with the same number of elements, then a better approximation for eigenvalues occurs. The geometry in Figure 6.1a and b gives better results.
- It is found that in both the cases the eigenvalues are converging with respect to the increasing number of elements.
- With reference to respective eigenvalues, the corresponding effective multiplication factors also converge.

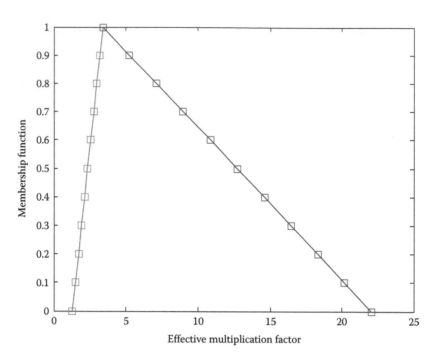

FIGURE 6.4
TFN for Figure 6.1b.

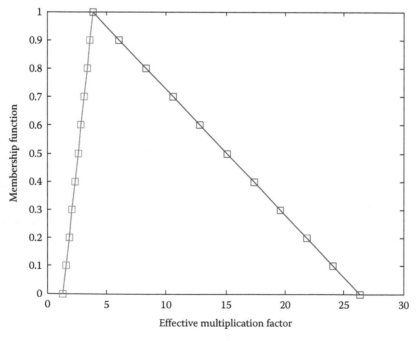

FIGURE 6.5
TFN for Figure 6.2a.

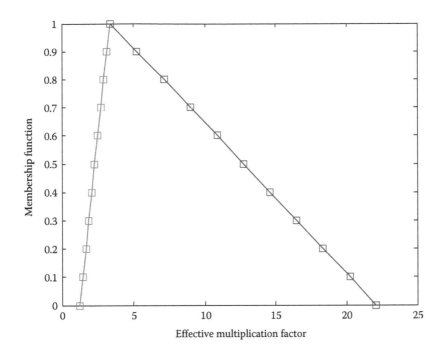

FIGURE 6.6
TFN for Figure 6.2b.

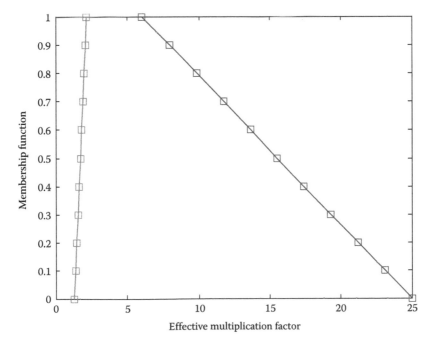

FIGURE 6.7
TRFN for Figure 6.1a.

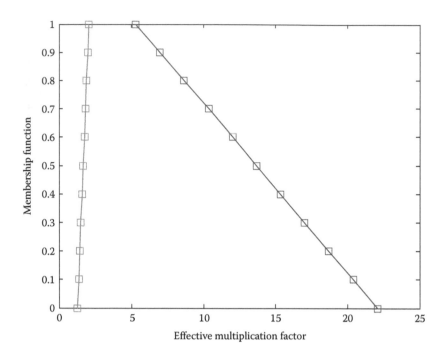

FIGURE 6.8
TRFN for Figure 6.1b.

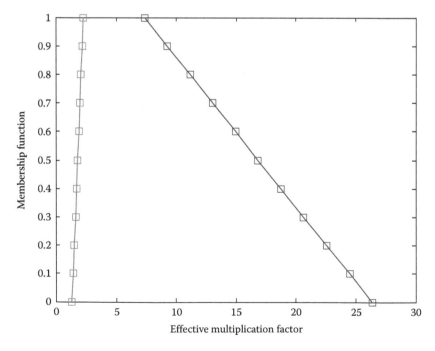

FIGURE 6.9
TRFN for Figure 6.2a.

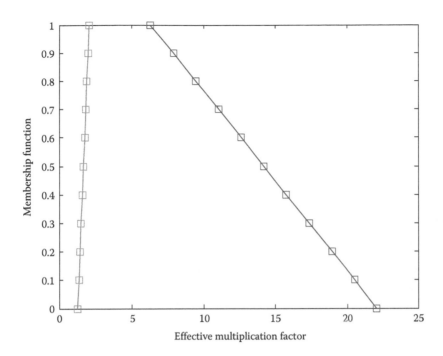

FIGURE 6.10
TRFN for Figure 6.2b.

In general, there is uncertainty in the system, and this uncertainty occurs due to the parameters involved, viz. the diffusion and absorption coefficients. To handle such an uncertainty, an alternate FFEM is presented here. Two different types of fuzzy numbers such as TFN and TRFN have been considered to investigate the uncertain eigenvalues and effective multiplication factors. As such,

- We get the different spectrum of eigenvalues for different discretizations (Figures 6.1a and b and 6.2a and b).
- When the parameters are taken as fuzzy, it is observed that the absorption coefficient is more sensitive.
- From Tables 6.2, 6.3, 6.5 and 6.6, it is found that when only the absorption coefficient is taken as fuzzy, then the uncertain bound for eigenvalues is wide in comparison with the case where only the diffusion coefficient is taken as fuzzy.
- When both the parameters are taken as fuzzy, then there is a much larger bound of uncertain eigenvalues than the previous cases. These results are depicted in Tables 6.4 and 6.7.
- The effective multiplication factors for the corresponding eigenvalues of Tables 6.4 and 6.7 are plotted in Figures 6.3 through 6.10.
- In view of Figures 6.3 through 6.10, we conclude that the uncertain bound of effective multiplication factors increases drastically.

- It is also seen that the geometry of the discretization plays a significant role. So, we may choose a better geometry to get a better distribution of uncertain effective multiplication factors.
- As already mentioned, depending upon the value of resulting uncertain effective multiplication factors, the neutron density fluctuates.

It may be noted that the reliability of the fuzzy results can be seen in the special cases, viz. crisp and interval, which are derived from the fuzzy values. As such, three cases are reported in this context.

6.4 Case Study 2

Here, we have considered a triangular (equilateral) bare reactor, having each side of 4 units, and it is discretized into a triangular element, as shown in Figure 6.11.

Fuzzy parameters are taken for the diffusion and absorption coefficients, which are presented in Table 6.8.

Initially, the governing one-group neutron diffusion equation is solved by considering only crisp parameters, and then the preceding method is used to solve the problem. Eigenvalues for both the crisp and fuzzy parameters are obtained, and the values are depicted in Table 6.9 for a different number of elements in the FEM and FFEM discretization.

For better visualization of the obtained results, eigenvalues for different numbers of discretizations of the domain are plotted and shown in Figures 6.12 through 6.19.

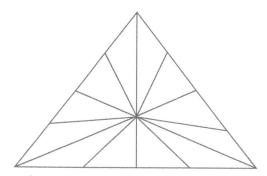

FIGURE 6.11
Triangular element discretization of triangular plate.

TABLE 6.8

Triangular Fuzzy Numbers for Uncertain Parameters

Parameters	Crisp Value	TFN
Diffusion coefficient	1	$[0.5+0.5\alpha, 1.5-0.5\alpha]$
Absorption coefficient	1	$[0.5+0.5\alpha, 1.5-0.5\alpha]$

TABLE 6.9

Crisp and Triangular Fuzzy Eigenvalues for Triangular Plate

Number of Elements	Crisp Eigenvalues	Triangular Fuzzy Eigenvalues
6	0.6425	[0.6377, 0.6425, 0.647]
12	0.6264	[0.6236, 0.6264, 0.6297]
24	0.526	[0.5251, 0.526, 0.527]
48	0.5083	[0.508, 0.5083, 0.5087]
96	0.5034	[0.5032, 0.5034, 0.5036]
192	0.5015	[0.5015, 0.5015, 0.5016]
384	0.5007	[0.5007, 0.5007, 0.5008]
1536	0.5002	[0.5002, 0.5002, 0.5002]

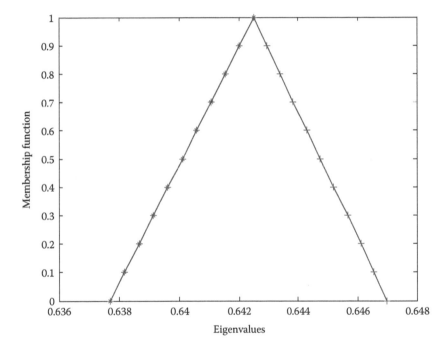

FIGURE 6.12
Six elements discretization of the domain.

The variation of both the crisp and fuzzy eigenvalues may be studied from Figure 6.20. Here, a set of eigenvalues are given and the convergence is studied.

6.4.1 Discussion

Here, a triangular bare reactor is considered and the neutron flux at the centre of the triangular geometry is taken as zero. The geometry is discretized into a number of triangular elements as given in Figure 6.11. So, the neutron flux distributions are studied for other nodal points. By solving the eigenvalue problem, we get a set of eigenvalues for a

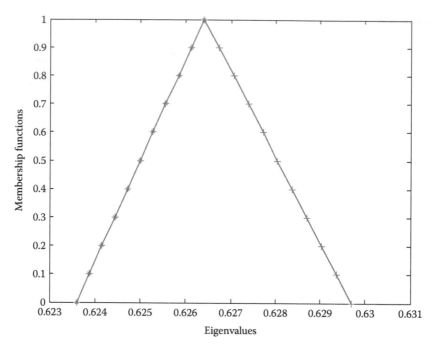

FIGURE 6.13
Twelve elements discretization of the domain.

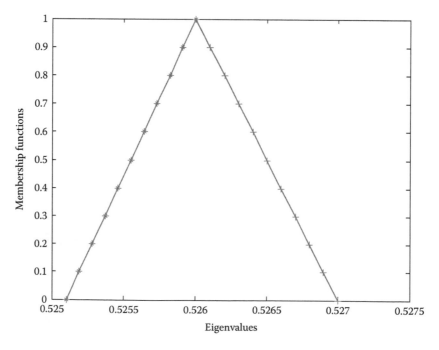

FIGURE 6.14
Twenty-four elements discretization of the domain.

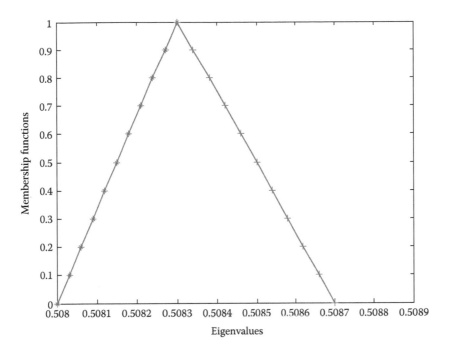

FIGURE 6.15
Forty-eight elements discretization of the domain.

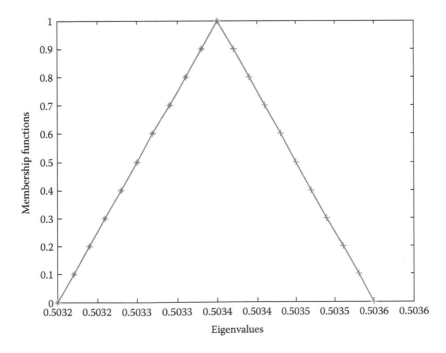

FIGURE 6.16
Ninety-six elements discretization of the domain.

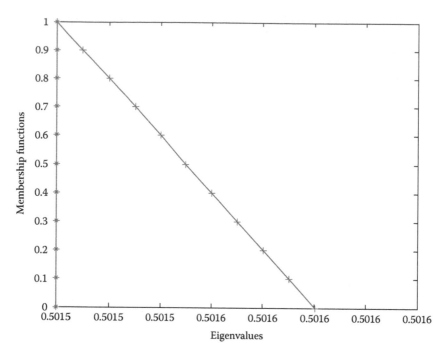

FIGURE 6.17
One hundred ninety-two elements discretization of the domain.

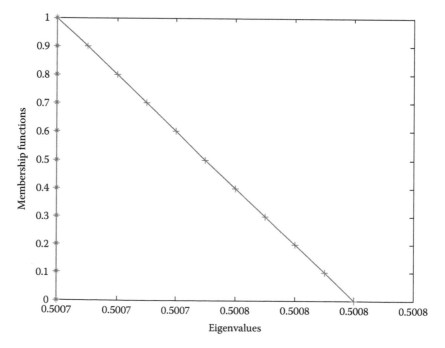

FIGURE 6.18
Three hundred eighty-four elements discretization of the domain.

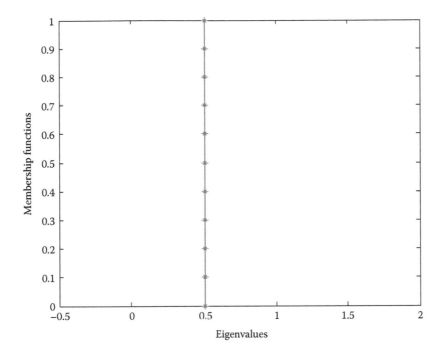

FIGURE 6.19
One thousand five-hundred thirty-six elements discretization of the domain.

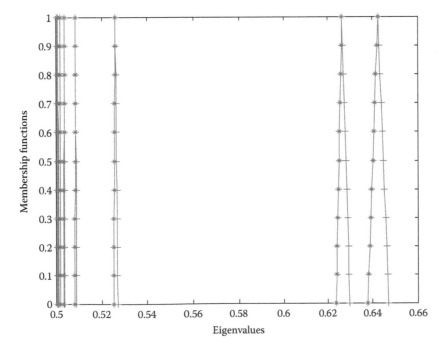

FIGURE 6.20
Triangular fuzzy membership functions for various discretizations of the domain.

different number of element discretizations. It is seen that the eigenvalues are converging with respect to the increase in the number of elements for the discretized triangular bare reactor. The pattern of the convergence is presented in Table 6.9.

As we increase the number of elements, the uncertain eigenvalues get converged. Furthermore, the uncertain width of eigenvalues decreases with an increase in number of discretizations of the triangular bare homogeneous reactor.

From the obtained results, it is observed that the shape of the TFN changes as we discretize the domain into elements, and these are shown in Figures 6.12 through 6.19. This variation of TFN occurs due to the left, right and centre values of the TFN. As we move on with the increase in the number of discretization of the domain, we get the right angular shaped fuzzy number. The trends of the shapes are shown in Figures 6.16 through 6.19. In Figures 6.17 and 6.18, the left and centre values are obtained as the same, so we get the left monotonically increasing function parallel to the membership functions axis, and the width of the left bound from the centre becomes zero. In Figure 6.19, the left, right and centre values approximately coincide and hence we get a straight line parallel to the membership function axis. Here, the variations of the eigenvalues become constant with the change of the membership functions. From Figure 6.20, it is seen that if we go on increasing the number of discretization of the domain, we get a series of uncertain eigenvalues and these triangular fuzzy eigenvalues converge into a constant value.

Finally, it may be noted that the reliability of the fuzzy results can be seen in the special cases, viz. crisp and interval, which are derived from the fuzzy values.

Bibliography

Azekura, K. 1980. New finite element solution technique for neutron diffusion equations. *Journal of Nuclear Science and Technology* 17:89–97.

Bart, M. N., Jose, A. E., Nico, S., Julio R. B. and Ashim, K. D. 2011. Fuzzy finite element analysis of heat conduction problems with uncertain parameters. *Journal of Food Engineering* 103:38–46.

Biswas, D., Ram, K. S. and Rao, S. S. 1976. Application of 'natural coordinate system' in the finite element solution of multigroup neutron diffusion equation. *Annals of Nuclear Energy* 3:465–469.

Cavdar, S. and Ozgener, H. A. 2004. A finite element/boundary element hybrid method for 2-D neutron diffusion calculations. *Annals of Nuclear Energy* 31:1555–1582.

Chakraverty, S. and Nayak, S. 2012. Fuzzy finite element method for solving uncertain heat conduction problems. *Coupled System Mechanics* 1(4):345–360.

Chakraverty, S. and Nayak, S. 2013. Non probabilistic solution of uncertain neutron diffusion equation for imprecisely defined homogeneous bare reactor. *Annals of Nuclear Energy* 62:251–259.

Dababneh, S., Khasawneh, K. and Odibat, Z. 2011. An alternative solution of the neutron diffusion equation in cylindrical symmetry. *Annals of Nuclear Energy* 38:1140–1143.

Glasstone, S. and Sesonke, A. 2004. *Nuclear Reactor Engineering*, 4th edn., Vol. 1. CBS Publishers and Distributors Private Limited, New Delhi, India.

Nagaya, Y., Chiba, G., Mori, T., Irwanto, D. and Nakajima, K. 2010. Comparison of Monte Carlo calculation methods for effective delayed neutron fraction. *Annals of Nuclear Energy* 37:1308–1315.

Nagaya, Y. and Mori, T. 2011. Calculation of effective delayed neutron fraction with Monte Carlo perturbation techniques. *Annals of Nuclear Energy* 38:254–260.

Nayak, S. and Chakraverty, S., 2013. Non-probabilistic approach to investigate uncertain conjugate heat transfer in an imprecisely defined plate. *International Journal of Heat and Mass Transfer* 67:445–454.

Nayak, S. and Chakraverty, S. 2015. Numerical solution of uncertain neutron diffusion equation for imprecisely defined homogeneous triangular bare reactor. *Sadhana* 40:2095–2109.

Rokrok, B., Minuchehr, H. and Zolfaghari, A. 2012. Element-free Galerkin modeling of neutron diffusion equation in *X–Y* geometry. *Annals of Nuclear Energy* 43:39–48.

Shi, B. and Petrovic, B. 2011. Implementation of the modified power iteration method to two-group Monte Carlo eigenvalue problems. *Annals of Nuclear Energy* 38:781–787.

Sjenitzer, B. L. and Hoogenboom, J. E. 2011. Variance reduction for fixed-source Monte Carlo calculations in multiplying systems by improving chain-length statistics. *Annals of Nuclear Energy* 38:2195–2203.

Yamamoto, T. 2012. Monte Carlo method with complex weights for neutron leakage-corrected calculations and anisotropic diffusion coefficient generations. *Annals of Nuclear Energy* 50:141–149.

Zadeh, L. A. 1965. Fuzzy sets. *Information and Control* 8:338–353.

7

Multigroup Model

Scattering neutrons produced by fission have a high range of energies. In a nuclear reactor, these neutrons are slowed down by scattering collisions with atomic nuclei until they are thermalized. In the thermal energy region, the neutrons exchange energy with the moderator atoms. Therefore, there is an up-scattering of neutrons such that neutrons gain energy, as well as the common down-scattering occurs and neutrons lose energy. As a result of various interactions, the neutron energies in a reactor core vary approximately from about 10 MeV to 0.001 eV. These energy ranges are divided into a finite number of discrete energy groups. Hence, we get multigroup neutron diffusion equations, which is the focus of this chapter. In the first section, some important information from the existing literature on the multigroup neutron diffusion equation is presented. Then, the formulation of the problem and a finite element procedure to solve these types of problems are explained in the second and third sections, respectively.

7.1 Background

Various authors have solved multigroup neutron diffusion equations with crisp parameters by considering numerical methods. In this regard, Ziver and Goddard (1981) used the finite element method to handle the multigroup diffusion transport equation. They have considered linear triangular and rectangular elements for the spatial dependence of the angular flux and spherical harmonic expansions for the angular dependence. Here, multigroup solutions are based on a maximum principle for the one-group second-order even-parity transport equation. Furthermore, Fletcher (1981) used the weighted residue method to solve the multigroup transport equation. The main aim is to enable high-order polynomial approximations to the flux. Then, the neutron transport equation is solved by expanding the flux in a series of non-normalized spherical harmonics, obtaining second-order diffusion-like equations for the coefficients in that series. In this context, a variational finite element method is used by Wood and De Oliveira (1984) to solve the multigroup neutron transport equation. They have taken a few examples and benchmark problems to demonstrate the obtained results.

De Oliveira (1986) solved multigroup neutron transport equations having anisotropic scattering of neutrons, and it has been investigated with the help of the variational finite element–spherical harmonic method. Wood (1986) has solved multigroup anisotropic scattering of neutrons by using the multigroup finite element code. Two test problems are given by the authors to demonstrate the codes. Riyait and Ackroyd (1987) considered a 1D slab to investigate the anisotropic scattering of neutrons using the finite element method. Further, Aydin and Atalay (2007) considered the steady-state two-group neutron diffusion equation for a critical reactor. They solved inverse neutron diffusion problems using a non-iterative direct approach, whereas Militão et al. (2012) used a numerical method, which is free of spatial truncation error and solved one-speed slab-geometry constant fixed-source adjoint discrete ordinates (SN) problems.

7.2 Group Diffusion Equation

As mentioned earlier, the energy ranges are divided into a finite number of discrete energy groups, which results in a multigroup equation. The division of the neutron energy range into G groups is shown in Figure 7.1. The maximum and minimum energies in the range of interest are denoted by E_0 and E_G, respectively. Here, the first group represents the group with the highest energy, that is $g = 1$. The g value increases with the decrease in neutron energy. Consider a neutron which is introduced into a group g' by fission or scattering. Then, the neutron will pass during moderation into a group g (i.e. at lower energy), where $g > g'$, whereas a certain amount of up-scattering may also occur (i.e. $g < g'$) in the thermal energy region.

In steady state, the neutron balance in any group g can be represented as follows:

$$\text{Leakage from group } g - \text{Absorption in group } g - \text{Scattering out of group } g$$
$$+ \text{Scattering into group } g + \text{Production in group } g = 0 \tag{7.1}$$

The parameters D_g (diffusion coefficient) and Σ_g (for absorption and scattering) are introduced here to derive the expression Equation 7.1 for the rates of these processes. The actual rates of neutron interactions within the group may be obtained by combining these parameters with the group flux ϕ_g.

Take a small volume element which is located at a point x. Here, for simplicity, a single coordinate is used but the results are applicable to any coordinate system. According to the diffusion theory, the leakage from group g (term 1) in the neutron balance Equation 7.1 can be represented by

$$\text{Term } 1 = -\nabla \cdot \left(D_g(x) \nabla \phi_g(x) \right) = -\frac{d}{dx}\left(D_g(x) \frac{d}{dx}\left(\phi_g(x)\right) \right), \tag{7.2}$$

where $D_g(x)$ and $\phi_g(x)$ are the group g diffusion coefficient and the neutron flux, respectively, at the point x in the system.

Absorption in group g (term 2) and scattering out of g (term 3) together represent the total rate of neutron interaction, which are absorption and scattering of neutrons in group g. The sum of these two terms is represented as

$$\text{Term } 2 + \text{Term } 3 = \Sigma_g^t(x)\phi(x), \tag{7.3}$$

where $\Sigma_g^t = \Sigma_g^a + \Sigma_g^s$ is the total interaction cross section for neutrons in group g. Scattering into group g (term 4) is defined as

$$\text{Term } 4 = \sum_{g'=1}^{G} \Sigma_{g' \to g}^s (x)\phi_{g'}(x) \tag{7.4}$$

FIGURE 7.1
Divisions of neutron into G groups.

where $\Sigma_{g'\to g}^s$ denotes the cross section for the scattering of neutrons from any group g' into g. This term also includes the scattering in which the neutron remains in the same group (when $g'=g$). The summation over values g' from 1 to G allows for scattering from all groups into group g.

Finally, there is a production in group g (term 5), which represents the rate at which neutrons are produced in the group. The term 5 can be written in the following way:

$$\text{Term } 5 = F_g(x). \tag{7.5}$$

Term 5 is equal to the rate at which neutrons with energies in group g are generated as a result of fission by the neutrons of all energies. Now, considering all the terms together, the steady-state neutron balance equation for group g can be written as

$$-\frac{d}{dx}\left(D_g(x)\frac{d}{dx}\big(\phi_g(x)\big)\right)-\Sigma_g^t(x)\phi(x)+\sum_{g'=1}^{G}\Sigma_{g'\to g}^s(x)\phi_{g'}(x)+F_g(x)=0. \tag{7.6}$$

Finally, the complete set of multigroup diffusion equations (7.6) consists of G equations with $g=1,2,\ldots,G$. These equations are coupled due to the flux ϕ_g in group g is dependent on the values of $\phi_{g'}$ in other groups.

7.3 Formulation

Let us consider standard coupled differential equations in their compact forms:

$$\frac{d^2\phi_1}{dx^2}=c_{11}\phi_1+c_{12}\phi_2+\cdots+c_{1n}\phi_n$$

$$\frac{d^2\phi_2}{dx^2}=c_{21}\phi_1+c_{22}\phi_2+\cdots+c_{2n}\phi_n \tag{7.7}$$

$$\vdots$$

$$\frac{d^2\phi_n}{dx^2}=c_{n1}\phi_1+c_{n2}\phi_2+\cdots+c_{nn}\phi_n$$

where $c_{ij}, i,j=1,2,\ldots,n$ and $\phi_i, i=1,2,\ldots,n$ are the coefficients and flux of the system.

To find the approximate flux (ϕ), Galarkin's weighted residue method has been used here. In this method, we need the residue equation, which can be obtained from Equation 7.7. Here, the residue equation is

$$\frac{d^2\phi_1}{dx^2}-c_{11}\phi_1-c_{12}\phi_2-\cdots-c_{1n}\phi_n=0$$

$$\frac{d^2\phi_2}{dx^2}-c_{21}\phi_1-c_{22}\phi_2-\cdots-c_{2n}\phi_n=0 \tag{7.8}$$

$$\vdots$$

$$\frac{d^2\phi_n}{dx^2}-c_{n1}\phi_1-c_{n2}\phi_2-\cdots-c_{nn}\phi_n=0.$$

Using linear interpolation (linear element discretization), the weight functions or the shape functions are obtained. In vector form, these functions will be

$$\begin{bmatrix} N_1 & N_2 \end{bmatrix}^T = \begin{bmatrix} 1 - \dfrac{x}{l} & \dfrac{x}{l} \end{bmatrix}^T,$$

where l is the length of each element of the domain. Then, the shape functions are multiplied with the residue equations and integrated over the domain.

If we consider only two equations having two dependent variables, then the residue will be

$$\left(\frac{d^2\phi_1}{dx^2} - c_{11}\phi_1 - c_{12}\phi_2 \right) = 0$$
$$\left(\frac{d^2\phi_2}{dx^2} - c_{21}\phi_1 - c_{22}\phi_2 \right) = 0, \quad i = 1, 2 \tag{7.9}$$

Now, Equation 7.9 is multiplied with the shape functions and it gives

$$N_i \left(\frac{d^2\phi_1}{dx^2} - c_{11}\phi_1 - c_{12}\phi_2 \right) = 0$$
$$N_i \left(\frac{d^2\phi_2}{dx^2} - c_{21}\phi_1 - c_{22}\phi_2 \right) = 0, \quad i = 1, 2 \tag{7.10}$$

Integrating Equation 7.10 over the domain Ω, we get

$$\int_\Omega N_i \left(\frac{d^2\phi_1}{dx^2} - c_{11}\phi_1 - c_{12}\phi_2 \right) = 0$$
$$\int_\Omega N_i \left(\frac{d^2\phi_2}{dx^2} - c_{21}\phi_1 - c_{22}\phi_2 \right) = 0, \quad i = 1, 2 \tag{7.11}$$

Now, adjusting the boundary condition Equation 7.11 can be solved (which is a set of algebraic equations) and unknowns are obtained. This concept is used to investigate the multigroup neutron diffusion equation, which is discussed in the following section.

7.4 Multigroup Neutron Diffusion Equation

Let us consider a 1D reactor core which is divided into various energy groups and different regions having constant material properties. The discretization of the reactor core into various groups is shown in Figure 7.1.

Using the previous procedure, the shape functions are multiplied with Equation 7.6, and we will have

$$[N_g] \frac{d}{dx}\left[D_g(x) \frac{d\phi_g(x)}{dx} \right] - [N_g] \sum_g {}^t(x)\phi_g(x) + [N_g] \sum_{g'=1}^G {}^s_{g' \to g}(x)\phi_{g'}(x) + [N_g]F_g(x) = 0. \tag{7.12}$$

Integrating Equation 7.11 over the domain, we get

$$\int_{\Omega}\left([N_g]\frac{d}{dx}\left[D_g(x)\frac{d\phi_g(x)}{dx}\right]-[N_g]\sum_g^t(x)\phi_g(x)+[N_g]\sum_{g'=1}^G\sum_{g'\rightarrow g}^s(x)\phi_{g'}(x)+[N_g]F_g(x)\right)dv=0$$

(7.13)

Further simplification of Equation 7.13 gives a system of algebraic equations. In matrix form, these algebraic equations look like

$$[K_g]\{\phi\}=[Q]$$ (7.14)

Here,
$[K_g]$ is the stiffness matrix for the coupled neutron diffusion equation
$\{\phi\}$ is the fuzzy neutron flux vector

In steady case $[Q]$, the matrix is zero.

Bibliography

Aydin, M. and Atalay, M. A. 2007. Inverse neutron diffusion problems in reactor design. *Journal of Nuclear Science and Technology* 44:1142–1148.

De Oliveira, C. R. E. 1986. An arbitrary geometry finite element method for multigroup neutron transport with anisotropic scattering. *Progress in Nuclear Energy* 18:227–236.

Fletcher, J. K. 1981. A solution of the multigroup transport equation using a weighted residual technique. *Annals of Nuclear Energy* 8:647–656.

Glasstone, S. and Sesonke, A. 2004. *Nuclear Reactor Engineering*, 4th edn., Vol. 1. CBS Publishers and Distributors Private Limited, New Delhi, India.

Militão, D. S., Filho, H. A. and Barros, R. C. 2012. A numerical method for monoenergetic slab-geometry fixed-source adjoint transport problems in the discrete ordinates formulation with no spatial truncation error. *International Journal of Nuclear Energy Science and Technology* 7:151–165.

Riyait, N. S. and Ackroyd, R. T. 1987. The finite element method for multigroup neutron transport: Anisotropic scattering in 1-D slab geometry. *Annals of Nuclear Energy* 14:113–133.

Sjenitzer, B. L. and Hoogenboom, J. E. 2013. Dynamic Monte Carlo method for nuclear reactor kinetics calculations. *Nuclear Science and Engineering* 175:94–107.

Wood, J. 1986. Multigroup anisotroping scattering in the finite element method. *Progress in Nuclear Energy* 18:91–100.

Wood, J. and De Oliveira, C. 1984. A multigroup finite element solution of the neutron transport equation—I: X–Y geometry. *Annals of Nuclear Energy* 11:229–243.

Ziver, A. K. and Goddard, A. J. H. 1981. A finite element method for multigroup diffusion-transport problems in two dimensions. *Annals of Nuclear Energy* 8:689–698.

8

Uncertain Multigroup Model

This chapter deals with the solution of the multigroup neutron diffusion equation under an uncertain environment. Here, the multigroup neutron diffusion equation for the steady-state case is considered, and a two-group neutron diffusion equation for an example problem is investigated by using the fuzzy finite element method. An example benchmark problem is demonstrated with uncertain parameters. Various parameters such as thermal conductivity, diffusion, group fission and neutron interaction constants are taken as fuzzy, and uncertain solutions, viz. thermal- and fast-group fluxes, are discussed.

8.1 Fuzzy Finite Element for Coupled Differential Equations

Let us consider the standard fuzzy coupled differential equations in the α-cut form as

$$
\begin{aligned}
\frac{d^2\left[\underline{\phi_1}(\alpha), \overline{\phi_1}(\alpha)\right]}{dx^2} &= \left[\underline{c_{11}}(\alpha), \overline{c_{11}}(\alpha)\right]\left[\underline{\phi_1}(\alpha), \overline{\phi_1}(\alpha)\right] + \left[\underline{c_{12}}(\alpha), \overline{c_{12}}(\alpha)\right]\left[\underline{\phi_2}(\alpha), \overline{\phi_2}(\alpha)\right] \\
\frac{d^2\left[\underline{\phi_2}(\alpha), \overline{\phi_2}(\alpha)\right]}{dx^2} &= \left[\underline{c_{21}}(\alpha), \overline{c_{21}}(\alpha)\right]\left[\underline{\phi_1}(\alpha), \overline{\phi_1}(\alpha)\right] + \left[\underline{c_{22}}(\alpha), \overline{c_{22}}(\alpha)\right]\left[\underline{\phi_2}(\alpha), \overline{\phi_2}(\alpha)\right]
\end{aligned}
\tag{8.1}
$$

Equation 8.1 may be written in compact form as follows:

$$
\begin{aligned}
\frac{d^2\tilde{\phi}_1}{dx^2} &= \tilde{c}_{11}\tilde{\phi}_1 + \tilde{c}_{12}\tilde{\phi}_2 \\
\frac{d^2\tilde{\phi}_2}{dx^2} &= \tilde{c}_{21}\tilde{\phi}_1 + \tilde{c}_{22}\tilde{\phi}_2
\end{aligned}
\tag{8.2}
$$

where
'~' represents the fuzzy numbers
$\tilde{c}_{ij}, i \cdot j = 1, 2$ and $\tilde{\phi}_i, i = 1, 2$ are the coefficients and flux in the system

To find the approximate uncertain flux (ϕ), Galarkin's weighted residue method has been used here. Considering the linear element discretization, the shape functions would be

$$\begin{bmatrix} \widetilde{N}_1 & \widetilde{N}_2 \end{bmatrix}^T = \begin{bmatrix} 1 - \dfrac{\tilde{x}}{l} & \dfrac{\tilde{x}}{l} \end{bmatrix}^T,$$ where l is the length of each element of the domain. Now, multiplying the shape functions with the residue of Equation 8.2, we get

$$\widetilde{N}_i \left(\frac{d^2 \tilde{\phi}_1}{dx^2} - \tilde{c}_{11} \tilde{\phi}_1 - \tilde{c}_{12} \tilde{\phi}_2 \right) = 0$$

$$\widetilde{N}_i \left(\frac{d^2 \tilde{\phi}_2}{dx^2} - \tilde{c}_{21} \tilde{\phi}_1 - \tilde{c}_{22} \tilde{\phi}_2 \right) = 0, \quad i = 1, 2$$

(8.3)

Integrating Equation 8.3 over the domain Ω, we get

$$\int_\Omega \widetilde{N}_i \left(\frac{d^2 \tilde{\phi}_1}{dx^2} - \tilde{c}_{11} \tilde{\phi}_1 - \tilde{c}_{12} \tilde{\phi}_2 \right) = 0$$

$$\int_\Omega \widetilde{N}_i \left(\frac{d^2 \tilde{\phi}_2}{dx^2} - \tilde{c}_{21} \tilde{\phi}_1 - \tilde{c}_{22} \tilde{\phi}_2 \right) = 0, \quad i = 1, 2$$

(8.4)

The solution of Equation 8.4 gives the uncertain neutron flux ($\tilde{\phi}$). This idea has been used for the formulation of multigroup neutron diffusion equations.

8.2 Fuzzy Multigroup Neutron Diffusion Equation

Let us consider the 1D reactor core, which is divided into various energy groups and different regions having constant material properties. The discretization of the reactor core into various groups is shown in Figure 7.1.

The general form of the fuzzy neutron diffusion equation may be written as (Glasstone and Sesonke 2004)

$$\frac{d}{dx}\left[\widetilde{D}_g(x) \frac{d\tilde{\phi}_g(x)}{dx} \right] - \sum_g^t (x) \tilde{\phi}_g(x) + \sum_{g'=1}^G \overset{s}{\underset{g' \to g}{\Sigma}} (x) \tilde{\phi}_{g'}(x) + \widetilde{F}_g(x) = 0$$

(8.5)

where
$g = 1, 2, \ldots G$
'~' denotes the fuzziness

Using this procedure, the shape functions are multiplied with Equation 8.5 and minimized to get

$$\begin{bmatrix} \widetilde{N}_g \end{bmatrix} \frac{d}{dx}\left[\widetilde{D}_g(x) \frac{d\tilde{\phi}_g(x)}{dx} \right] - \begin{bmatrix} \widetilde{N}_g \end{bmatrix} \sum_g^t (x) \tilde{\phi}_g(x) + \begin{bmatrix} \widetilde{N}_g \end{bmatrix} \sum_{g'=1}^G \overset{s}{\underset{g' \to g}{\Sigma}} (x) \tilde{\phi}_{g'}(x) + \begin{bmatrix} \widetilde{N}_g \end{bmatrix} \widetilde{F}_g(x) = 0$$

(8.6)

Integrating Equation 8.6 over the domain, we get

$$\int_{\Omega} \left(\left[\widetilde{N}_g \right] \frac{d}{dx} \left[\widetilde{D}_g(x) \frac{d\tilde{\phi}_g(x)}{dx} \right] - \left[\widetilde{N}_g \right] \sum_g^t (x) \tilde{\phi}_g(x) + \left[\widetilde{N}_g \right] \sum_{g'=1}^G \mathop{\Sigma}\limits_{g' \to g}^s (x) \tilde{\phi}_{g'}(x) + \left[\widetilde{N}_g \right] \widetilde{F}_g(x) \right) dv = 0$$

(8.7)

Further simplification of Equation 8.7 gives us a system of algebraic equations. In matrix form, this algebraic equations look like

$$\left[\widetilde{K}_g \right] \{ \tilde{\phi} \} = \left[\widetilde{Q} \right]$$

(8.8)

Here,

$\left[\widetilde{K}_g \right]$ is the fuzzy stiffness matrix for the coupled fuzzy neutron diffusion equation

$\{ \tilde{\phi} \}$ is the fuzzy neutron flux vector

In steady case, the $\left[\widetilde{Q} \right]$ matrix is zero. Equation 8.8 is a fully fuzzy system of equations, which is tedious to handle. But this difficulty may be overcome by using the method discussed in Nayak and Chakraverty (2013).

Let us consider the two-group fuzzy neutron diffusion equation (Wood and De Oliveira 1984)

$$-\frac{d^2\tilde{\phi}_1}{dx^2} = \tilde{m}_{11}\tilde{\phi}_1 + \tilde{m}_{12}\tilde{\phi}_2$$

$$-\frac{d^2\tilde{\phi}_2}{dx^2} = \tilde{m}_{21}\tilde{\phi}_1 + \tilde{m}_{22}\tilde{\phi}_2$$

(8.9)

where

$$\tilde{m}_{11} = \frac{\left(\tilde{v} \dfrac{\widetilde{\Sigma}_{f1}}{k} - \widetilde{\Sigma}_{r1} \right)}{\widetilde{D}_1},$$

$$\tilde{m}_{12} = \frac{\tilde{v}\widetilde{\Sigma}_{f2}}{\widetilde{D}_1 k},$$

$$\tilde{m}_{21} = \frac{\widetilde{\Sigma}_{12}}{\widetilde{D}_2},$$

$$\tilde{m}_{22} = -\frac{\widetilde{\Sigma}_{a2}}{\widetilde{D}_2}.$$

Using the Galarkin fuzzy finite element formulation for each element having length l, we get the following stiffness matrix:

$$
\begin{bmatrix}
\dfrac{2}{l} + \widetilde{m}_{11}\dfrac{l}{3} & \widetilde{m}_{12}\dfrac{l}{3} & -\dfrac{2}{l} + \widetilde{m}_{11}\left(\dfrac{l}{2} - \dfrac{l}{3}\right) & \widetilde{m}_{12}\left(\dfrac{l}{2} - \dfrac{l}{3}\right) \\[2ex]
\widetilde{m}_{21}\dfrac{l}{3} & \dfrac{2}{l} + \widetilde{m}_{21}\dfrac{l}{3} & \widetilde{m}_{21}\left(\dfrac{l}{2} - \dfrac{l}{3}\right) & -\dfrac{2}{l} + \widetilde{m}_{11}\left(\dfrac{l}{2} - \dfrac{l}{3}\right) \\[2ex]
\widetilde{m}_{11}\left(\dfrac{l}{2} - \dfrac{l}{3}\right) & \widetilde{m}_{12}\left(\dfrac{l}{2} - \dfrac{l}{3}\right) & \widetilde{m}_{11}\dfrac{l}{3} & \widetilde{m}_{12}\dfrac{l}{3} \\[2ex]
\widetilde{m}_{21}\left(\dfrac{l}{2} - \dfrac{l}{3}\right) & \widetilde{m}_{22}\left(\dfrac{l}{2} - \dfrac{l}{3}\right) & \widetilde{m}_{21}\dfrac{l}{3} & \widetilde{m}_{22}\dfrac{l}{3}
\end{bmatrix}
$$

where $\widetilde{m}_{ij}$, $i, j = 1, 2$ are fuzzy numbers.

8.3 Case Study

Consider a benchmark problem (ANL-BSS-6-A2) for the steady-state case. The core geometry is one dimensional and its length is considered to be 240 cm, which is divided into three regions, viz. 40, 160 and 40 cm, respectively (Wood and De Oliveira 1984). The boundary flux is zero, which is shown in Figure 8.1. The parameters used in this problem are encrypted in Table 8.1.

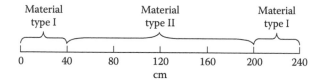

FIGURE 8.1
Geometry of the domain in one dimension.

TABLE 8.1

Crisp and Fuzzy Parameters for Different Regions of the Domain

Parameters	Region 1, 3		Region 2	
	Crisp	Triangular Fuzzy Number (TFN)	Crisp	Triangular Fuzzy Number (TFN)
D_1	1.5	[1, 1.5, 2]	1	[0.5, 1, 1.5]
D_2	0.5	[0.2, 0.5, 0.8]	0.5	[0.2, 0.5, 0.8]
Σ_{r1}	0.26	[0.2, 0.26, 0.32]	0.2	[0.1, 0.2, 0.3]
Σ_{a2}	0.18	[0.15, 0.18, 0.21]	0.08	[0.03, 0.08, 0.13]
Σ_{12}	0.015	[0.01, 0.015, 0.02]	0.01	[0.005, 0.01, 0.15]
$v\Sigma_{f1}$	0.01	[0.005, 0.01, 0.15]	0.005	[0.001, 0.005, 0.009]
$v\Sigma_{f2}$	0.2	[0.15, 0.2, 0.25]	0.099	[0.05, 0.099, 0.148]
k	1	1	1	1

Initially, this problem is solved for crisp parameters, and then the uncertain variations of parameters are handled. The traditional Galarkin finite element method has been used to investigate the problem with crisp parameters, and the obtained neutron fluxes are depicted in Figures 8.2 and 8.3.

Obtained uncertain neutron fluxes (when group fission constants, diffusion and neutron interaction coefficients as fuzzy) are shown in Figures 8.4 and 8.5. In these figures, uncertain thermal- and fast-group fluxes are graphically presented.

For different values of α, we get different interval values for the uncertain parameters. The variation of flux is graphically shown in Figures 8.6 and 8.7, respectively, when all the parameters are fuzzy and α is zero.

In order to see the sensitiveness of the parameters, we have considered different cases, viz. (1) diffusion coefficients (D_1, D_2), (2) neutron interaction coefficients (Σ_{r1}, Σ_{a2}, Σ_{12}) and (3) group fission constants ($v\Sigma_{f1}$, $v\Sigma_{f2}$) as fuzzy. The uncertain neutron fluxes for these cases are presented in Figures 8.8 through 8.13.

8.4 Results and Discussion

As mentioned earlier, the previous problem is investigated first for crisp parameters, and then uncertainties are considered to demonstrate the fuzzy finite element method (Nayak and Chakraverty 2013). The sensitivity of the uncertain parameters is analyzed by considering left, right and centre values for the obtained neutron flux. Some of the major issues are discussed as follows.

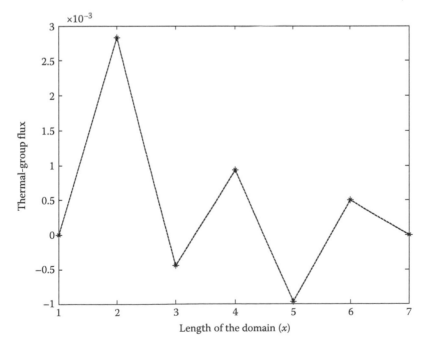

FIGURE 8.2
Thermal-group flux along the domain (with crisp parameters).

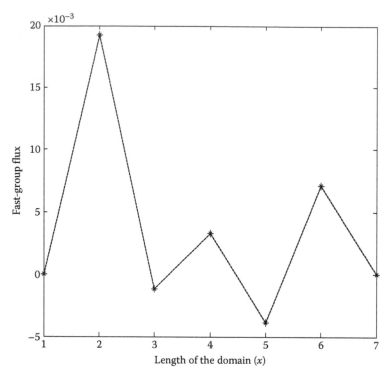

FIGURE 8.3
Fast-group flux along the domain (with crisp parameters).

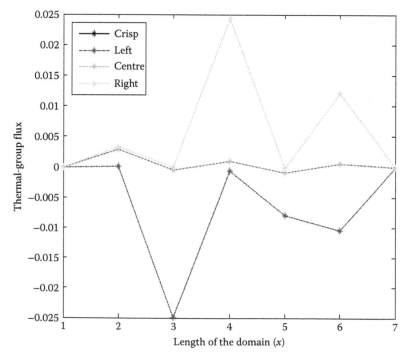

FIGURE 8.4
Thermal-group flux along the domain when all the parameters are fuzzy.

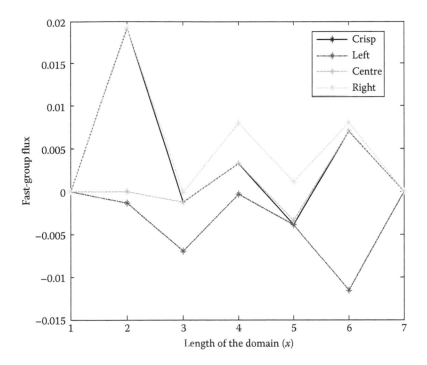

FIGURE 8.5
Fast-group flux along the domain when all the parameters are fuzzy.

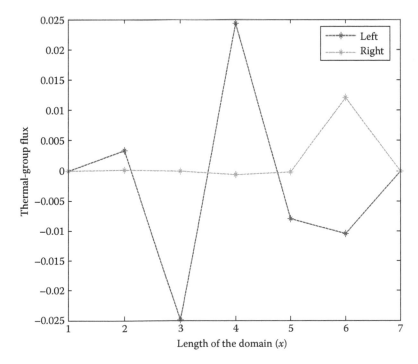

FIGURE 8.6
Thermal-group flux at α is zero when all the parameters are fuzzy.

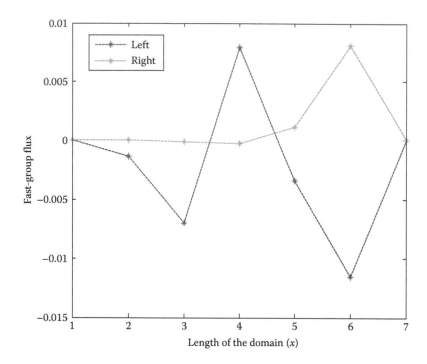

FIGURE 8.7
Fast-group flux at α is zero when all the parameters are fuzzy.

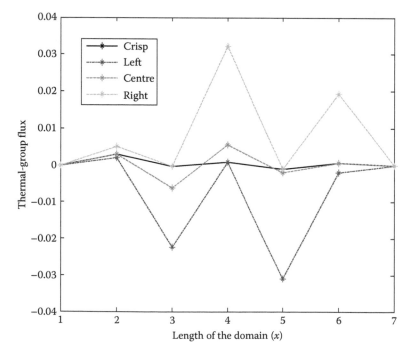

FIGURE 8.8
Thermal-group flux when D_1, D_2 are fuzzy.

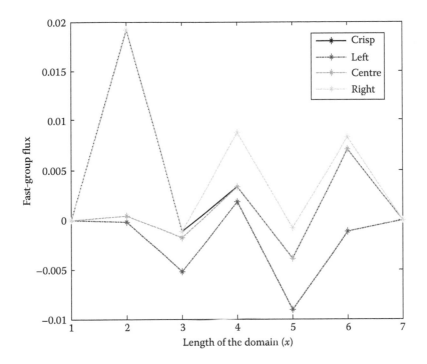

FIGURE 8.9
Fast-group flux when D_1, D_2 are fuzzy.

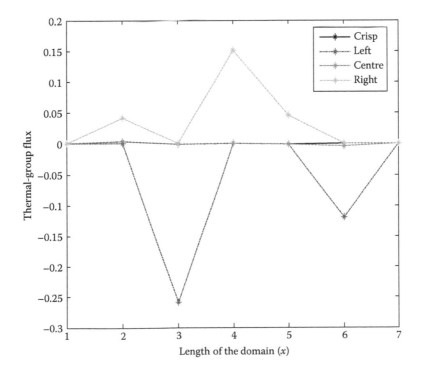

FIGURE 8.10
Thermal-group flux when $\Sigma_{r1}, \Sigma_{a2}, \Sigma_{12}$ are fuzzy.

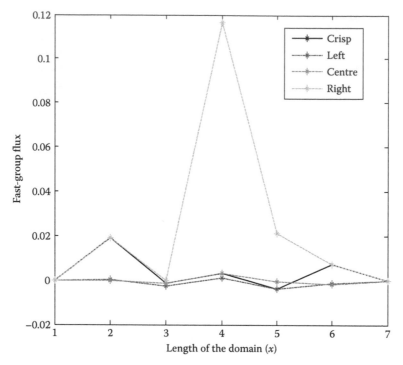

FIGURE 8.11
Fast-group flux when Σ_{r1}, Σ_{a2}, Σ_{12} are fuzzy.

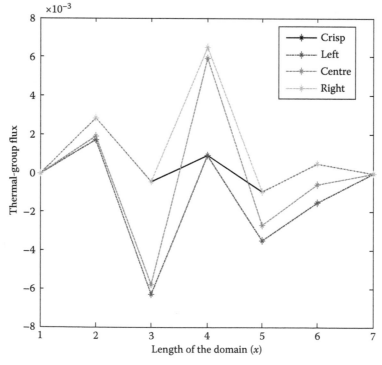

FIGURE 8.12
Thermal-group flux when $v\Sigma_{f1}$, $v\Sigma_{f2}$ are fuzzy.

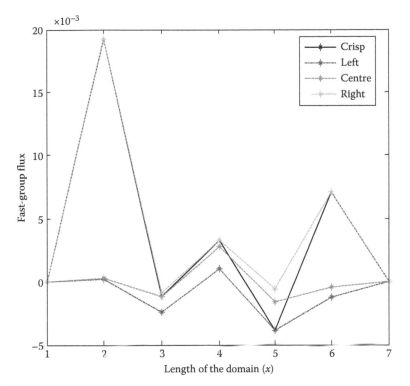

FIGURE 8.13
Fast-group flux when $v\Sigma_{f1}$, $v\Sigma_{f2}$ are fuzzy.

8.4.1 Thermal-Group Neutron Fluxes

When only the diffusion coefficients (D_1, D_2) are fuzzy and other parameters are crisp, then in the region 0–40 cm, left, centre and crisp values are close to each other. From Figure 8.8, it is noticed that there is a deviation of the centre and crisp values from 40 to 160 cm. After that, the centre and crisp values converge. We see that there is a sudden deviation (increase in the uncertain width) of right values in 80–160 cm and 160–240 cm. Similarly, there is sudden deviation of left values in 40–120 cm and 120–200 cm.

In the case where only the neutron interaction coefficients (Σ_{r1}, Σ_{a2}, Σ_{12}) are fuzzy, the distribution of uncertain thermal-group flux is shown in Figure 8.10. It is seen that the left, centre and crisp values are the same in the region 0–40 cm. From 40 to 120 cm, there is a sudden deviation of the left value. Then, from 120 to 160 cm the values are the same, and then there is a deviation. Similarly, the right values show deviation throughout the domain except in the region 200–240 cm, where the right, centre and crisp values are the same. Finally, it may be seen that the distribution of centre and crisp thermal-group fluxes is almost the same along the domain.

Next, we have considered the group fission constants ($v\Sigma_{f1}$, $v\Sigma_{f2}$) as only fuzzy, and the uncertain thermal-group fluxes are depicted in Figure 8.12. Here, the centre and right values are the same in 0–80 cm and 160–240 cm. The centre and left values are almost the same in 0–80 cm. In Figure 8.14, we see a continuous fluctuation of the centre and crisp values throughout the domain.

Finally, all the parameters are taken as fuzzy and the obtained results are depicted in Figure 8.4, where the crisp and centre values are very close. There is a sudden left and right deviation between 40–120 cm and 80–160 cm, respectively. From 0 to 80 cm, the right, centre and crisp values almost overlap.

Further, to check the sensitiveness of the used parameters, the deviation of left and right values from the centre and crisp values is shown graphically in Figures 8.14 through 8.17. In view of these four figures, it is concluded that when the group fission constants are fuzzy, the system is sensitive. A slight change in the parameter drastically changes the distribution of thermal-group fluxes. The width of the uncertain thermal-group fluxes increases with larger width values of Σ_{r1}, Σ_{a2}, Σ_{12}.

8.4.2 Fast-Group Neutron Fluxes

In Figure 8.9, we have considered only the diffusion coefficients (D_1, D_2) as fuzzy. Here, we observed that the distribution of the centre and crisp fast-group flux is similar to the case where all the parameters are taken as fuzzy, whereas the uncertain width maintains a consistency from 80 to 160 cm and there is a deviation of the left value distribution in the region 200 to 240 cm.

Next, we consider only the neutron interaction coefficients $(\Sigma_{r1}, \Sigma_{a2}, \Sigma_{12})$ as fuzzy and the obtained results are presented in Figure 8.11. Here, we may see that in the region 0–80 cm, the centre and left values are the same, and in the same fashion, the crisp and right distributions of fluxes are same. It is noticed that there is a sharp deviation of the right value from 80 to 200 cm. Finally, from 200 to 240 cm, the crisp and right distributions are the same.

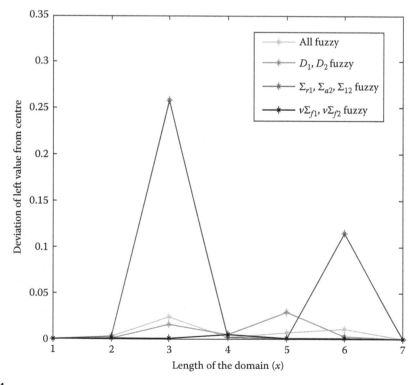

FIGURE 8.14

Left value of the thermal-group flux for different sub-cases with the centre value of the uncertain neutron fluxes when all parameters are fuzzy.

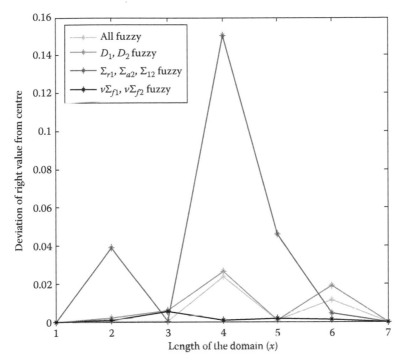

FIGURE 8.15

Right value of the thermal-group flux for different sub-cases with the centre value of the uncertain neutron fluxes when all parameters are fuzzy.

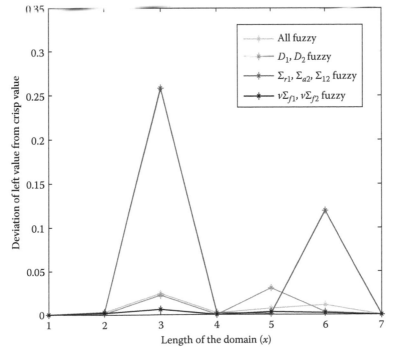

FIGURE 8.16

Left value of the uncertain thermal-group flux with crisp value.

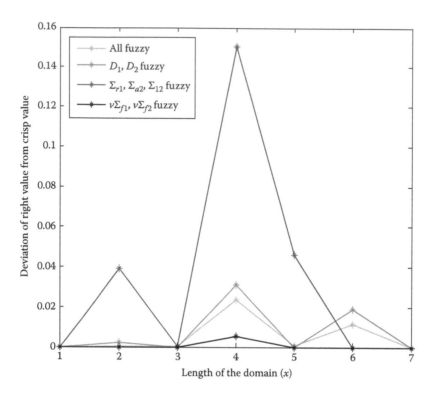

FIGURE 8.17
Right value of the uncertain thermal-group flux with the crisp value.

When the group fission constants ($v\Sigma_{f1}$, $v\Sigma_{f2}$) are taken as fuzzy, we observed that the right and crisp distributions are the same from 0 to 120 cm and from 200 to 240 cm. Further, it is investigated that the width of the uncertainty decreases in the region 80–160 cm, which may be visualized in Figure 8.13.

Finally, when all the parameters are taken as fuzzy, Figure 8.5 shows that in the region 0–80 cm, the right and crisp values of the fast-group flux are almost the same. Also, from 80 to 240 cm, the centre and crisp distribution of fluxes are the same. It is seen that there is a constant uncertain width of the fluxes from 80 to 160 cm, whereas the uncertain width increases widely from 160 to 200 cm and then decreases.

To investigate the sensitiveness of the parameters used, the deviation of the left and right flux distribution of fast group with the centre and crisp fluxes is depicted graphically in Figures 8.18 through 8.21. It is seen that in both the cases, the case where neutron interaction coefficients (Σ_{r1}, Σ_{a2}, Σ_{12}) are fuzzy, the system becomes sensitive. Here, an increase in the width of uncertainty in these parameters results in a drastic increase in the width of the fast-group fluxes.

From this analysis, it is seen that when neutron interaction coefficients are fuzzy then the uncertain thermal-group and fast-group fluxes are more sensitive than the other cases. Here, a small change in the values of neutron interaction coefficients affects more to the distribution of neutron flux and the error or width of uncertainty increases comparing with the other cases.

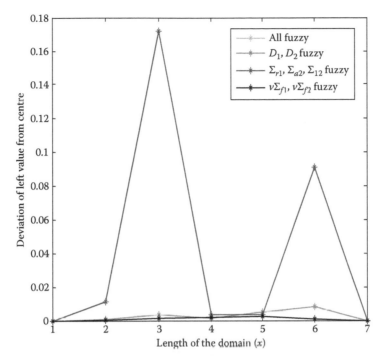

FIGURE 8.18
Left value of the fast-group flux for different sub-cases with the centre value of the uncertain neutron fluxes when all parameters are fuzzy.

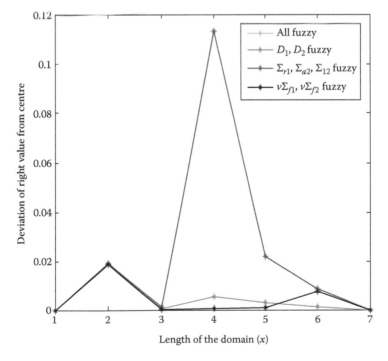

FIGURE 8.19
Right value of the fast-group flux for different sub-cases with the centre value of the uncertain neutron fluxes when all parameters are fuzzy.

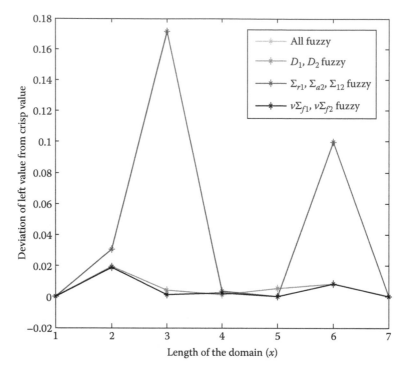

FIGURE 8.20
Left value of the uncertain fast-group flux with crisp value.

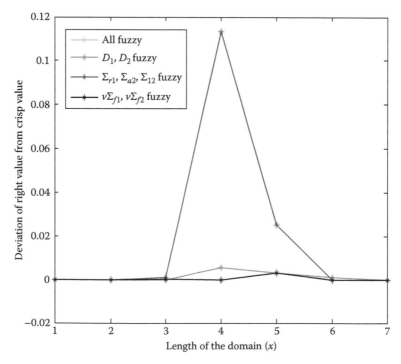

FIGURE 8.21
Right value of the uncertain fast-group flux with crisp value.

Bibliography

Aydin, M. and Atalay, M. A. 2007. Inverse neutron diffusion problems in reactor design. *Journal of Nuclear Science and Technology* 44(9):1142–1148.

Chakraverty, S. and Nayak, S. 2013a. Fuzzy finite element method in diffusion problems. In: *Mathematics of Uncertainty Modelling in the Analysis of Engineering and Science Problems*. IGI Global. pp. 309–328.

Chakraverty, S. and Nayak, S. 2013b. Non probabilistic solution of uncertain neutron diffusion equation for imprecisely defined homogeneous bare reactor. *Annals of Nuclear Energy* 62:251–259.

De Oliveira, C. R. E. 1986. An arbitrary geometry finite element method for multigroup neutron transport with anisotropic scattering. *Progress in Nuclear Energy* 18:227–236.

Fletcher, J. K. 1981. A solution of the multigroup transport equation using a weighted residual technique. *Annals of Nuclear Energy* 8:647–656.

Glasstone, S. and Sesonke, A. 2004. *Nuclear Reactor Engineering*, 4th edn., Vol. 1. CBS Publishers and Distributors Private Limited, New Delhi, India.

Militão, D. S., Filho, H. A. and Barros, R. C. 2012. A numerical method for monoenergetic slab-geometry fixed-source adjoint transport problems in the discrete ordinates formulation with no spatial truncation error. *International Journal of Nuclear Energy Science and Technology* 7(2):151–165.

Nayak, S. and Chakraverty, S. 2013. Non-probabilistic approach to investigate uncertain conjugate heat transfer in an imprecisely defined plate. *International Journal of Heat and Mass Transfer* 67:445–454.

Park, H. J., Shim, H. J., Joo, H. G. and Kim, C. H. 2013. Uncertainty quantification of few-group diffusion theory constants generated by the B_1 theory-augmented Monte Carlo method. *Nuclear Science and Engineering* 175:28–43.

Riyait, N. S. and Ackroyd, R. T. 1987. The finite element method for multigroup neutron transport: Anisotropic scattering in 1-D slab geometry. *Annals of Nuclear Energy* 14(3):113–133.

Sjenitzer, B. L. and Hoogenboom, J. E. 2013. Dynamic Monte Carlo method for nuclear reactor kinetics calculations. *Nuclear Science and Engineering* 175:94–107.

Wood, J. 1986. Multigroup anisotroping scattering in the finite element method. *Progress in Nuclear Energy* 18:91–100.

Wood, J. and De Oliveira, C. 1984. A multigroup finite element solution of the neutron transport equation-I: *X–Y* geometry. *Annals of Nuclear Energy* 11(5):229–243.

Zadeh, L. A. 1965. Fuzzy sets. *Information Control* 8:338–353.

Ziver, A. K. and Goddard, A. J. H. 1981. A finite element method for multigroup diffusion-transport problems in two dimensions. *Annals of Nuclear Energy* 8:689–698.

9

Point Kinetic Diffusion

The stability in the operation and controlling of nuclear power reactors are practical cases, which are to be performed in a safe manner. Accordingly, the kinetic concept is needed to study the transient changes from the departure of the reactor condition to the critical state. These situations occur under the conditions of (1) start-up, (2) shutdown and (3) accidental disturbances in a nuclear reactor.

In a reactor, the transient production of power plays an important role for determining the degree of damage that can result from an accident. The time-dependent power production depends on the effective multiplication factor k_{eff}, prompt and delayed neutron properties of the reactor kinetic equations. The neutron flux distribution is ignored in favour of an emphasis on its time behaviour. The reactor is viewed as a point and hence the term 'point reactor kinetics'. In this regard, a distinction must be made between the behaviours of the prompt and delayed neutrons.

The time-dependent diffusion equation states that the rate of change of the number of neutrons in a unit volume is just equal to the difference of the absorption rate of neutrons from the production rate of neutrons in the unit volume, which may be represented as (Glasstone and Sesonke 2004)

$$\frac{1}{v}\frac{\partial \phi(\overline{r},t)}{\partial t} = S(\overline{r},t) + D\nabla^2 \phi(\overline{r},t) - \Sigma_a \phi(\overline{r},t) \tag{9.1}$$

where
  D is the diffusion coefficient
  Σ_a is the absorption coefficient
  S is the source

9.1 Reactivity

A critical reactor has an effective multiplication factor k_{eff}, which is equal to unity. The effective multiplication factor can be either larger or less than unity, depending on the deviation of a nuclear reactor from its criticality. As such, it has an 'excess multiplication factor'.

$$k_{ex} = k_{eff} - 1 \tag{9.2}$$

which can be positive or negative.

The ratio of the excess multiplication factor to the effective multiplication factor is called 'excess reactivity' or 'reactivity' and is defined as

$$\rho = \frac{k_{ex}}{k_{eff}} = \frac{k_{eff} - 1}{k_{ex}} \tag{9.3}$$

The reactivity explains the deviation of the effective multiplication factor from unity under time-dependent conditions. In a steady-state case, the reactivity is zero.

If the deviation is small from criticality, which is by small deviations in temperature or voids during the normal operation, then the reactivity can be expressed as follows:

$$\rho \approx k_{ex} \approx k_{eff} - 1 \qquad (9.4)$$

In terms of reactivity, the effective multiplication factor may be expressed as

$$k_{eff} = \frac{1}{1-\rho} \qquad (9.5)$$

and the excess reactivity can be given as

$$k_{ex} = \frac{\rho}{1-\rho} \qquad (9.6)$$

Changes in the system's temperature, pressure or load may result the reactivity to be short-lived. It may also develop over a long period of time because of the fuel burn-up and the accumulation of fission products.

To hold the reactor power constant, means are devised to keep its reactivity constant such as the control rods or the chemical shim (which is a neutron absorber in the coolant or moderator).

9.2 Prompt and Delayed Neutrons

The neutrons emitted within a short interval of 10^{-17} s of the fission process are called 'prompt neutrons', which is about 99% of the fission neutrons. The remaining neutrons are delayed in their emission in the fission process itself and are known as 'delayed neutrons' (Hetrick 1971, Lamarsh 1983, Stacey Weston 2007).

For an example, we can consider the delayed neutron emission from the fission product, isotope Br^{87}, which has a half-life of 55.6 s. The beta decay of Br^{87}, through its two main branches of 2.6 and 8 MeV electrons, leads to the formation of Kr^{87} in its ground state, and it subsequently decays through two successive beta emissions into the stable isotope of Sr^{87}, whereas it is possible for the delayed neutron precursor Br^{87} nucleus to beta decay into an excited state of the Kr^{87} nucleus at an energy of 5.5 MeV, which is larger than the binding energy of a neutron in the Kr^{87} nucleus. In this case, the beta emission is followed by a neutron emission leading to the stable Kr^{87} isotope.

The fraction of delayed neutrons from U^{235} is only $\beta = 0.0065$. But it is smaller for Pu^{239} at $\beta = 0.0021$. The occurrence of delayed neutrons is crucial for the control of nuclear reactors. Even though the fraction of delayed neutrons is small, their presence provides a long time constant. Hence, it slows down the dynamic time response of a nuclear reactor to make it controllable by the withdrawal and insertion of control rods, containing nuclear absorbing materials such as boron.

The weighted average of the mean lifetime of the delayed neutrons is much larger than that of the prompt neutrons.

9.2.1 Delayed Neutron Parameters

Consider N number of delayed neutron groups (usually taken as 6), then the number of delayed neutrons produced per unit volume in the steady state is (Ghoshal 2010)

$$n_d = \sum \lambda_i C_i (\bar{r}) \text{ neutrons} / (\text{cm}^3 \cdot \text{s}) \tag{9.7}$$

where $C_i (\bar{r})$ and λ_i are the ith concentration of the beta emitter, which is a precursor of a delayed neutron emitter and a decay constant for the neutron emitter of the ith type, respectively.

For steady state, the generated fission products from the fission process are equal to those decaying through radioactive decay, which can be written as

$$w_i \Sigma_f \phi(\bar{r}) - \lambda_i C_i (\bar{r}) = 0, \quad i = 1, 2, 3, \ldots, N \tag{9.8}$$

where w_i is the fraction of fissions, which yields precursors of the ith type.

The total number of prompt and delayed neutrons produced per unit volume per unit time is represented as

$$n_t = n_d + (1 - \beta) \nu \Sigma_f \varepsilon \phi(\bar{r})$$
$$= \sum_{i=1}^{N} \lambda_i C_i (\bar{r}) + (1 - \beta) \nu \Sigma_f \varepsilon \phi(\bar{r}) \tag{9.9}$$

where β is the delayed neutrons fraction from fissions.

This β is related to the fraction of neutrons from the fission that is produced by the ith delayed neutron precursor group given by

$$\beta = \sum_{i=1}^{N} \beta_i \tag{9.10}$$

9.3 Point Reactor Kinetic Equations for a Non-Stationary One-Group Bare Reactor with Delayed Neutrons

In this case, the reactor is not critical and the effective multiplication factor is different from unity, and it is written as

$$k_{eff} = k_\infty l_{th} l_f$$
$$= \eta \varepsilon p f \frac{1}{(1 + L^2 B^2)} \frac{1}{(1 + \tau B^2)}$$
$$= \frac{\nu \Sigma_f}{\Sigma_{a\,fuel}} \varepsilon p \frac{\Sigma_{a\,fuel}}{\Sigma_a} \frac{1}{(1 + L^2 B^2)} \frac{1}{(1 + \tau B^2)}$$
$$= \frac{\nu \Sigma_f}{\Sigma_a} \varepsilon p \frac{1}{(1 + L^2 B^2)} \frac{1}{(1 + \tau B^2)}$$

where
$\Sigma_{a\,fuel}$ is the macroscopic fuel absorption cross section
Σ_a is the absorption cross section
B^2 is the geometrical buckling

The equations governing the precursors' concentrations will be a modification of Equation 9.9, accounting for the time dependence as

$$\frac{\partial C_i(\bar{r},t)}{\partial t} = w_i \Sigma_f \phi(\bar{r},t) - \lambda_i C_i(\bar{r},t), \quad i = 1,2,3,\ldots,N \tag{9.11}$$

The average energies of the delayed neutrons range are from (about) 0.25 to 0.62 MeV. The balance equation for the thermal neutrons in terms of the flux with a source term is

$$S(\bar{r},t) = (1-\beta) v\Sigma_f \varepsilon \phi(\bar{r},t) pl_f + p\varepsilon l_f \sum_{i=1}^{N} \lambda_i C_i(\bar{r},t) \tag{9.12}$$

Substituting Equation 9.12 into Equation 9.1 yields

$$\frac{1}{v}\frac{\partial \phi(\bar{r},t)}{\partial t} = (1-\beta) v\Sigma_f \varepsilon \phi(\bar{r},t) pl_f + p\varepsilon l_f \sum_{i=1}^{N} \lambda_i C_i(\bar{r},t) + D\nabla^2 \phi(\bar{r},t) - \Sigma_a \phi(\bar{r},t) \tag{9.13}$$

It is reasonable to suppose that the spatial variation of the concentration of the delayed neutron precursors is proportional to that of the neutron flux and that this mode persists even though the magnitude of the flux changes with time. Thus, let us assume

$$\phi(\bar{r},t) \equiv F(\bar{r})\phi(t)$$
$$C_i(\bar{r},t) \equiv F(\bar{r})C_i(t) \tag{9.14}$$

where

$$\nabla^2 F(\bar{r}) + B^2 F(\bar{r}) = 0 \tag{9.15}$$

and the boundary condition at the extrapolated radius of the reactor is $F(R_{extrapolated}) = 0$.
   Hence, Equation 9.13 becomes

$$\frac{1}{v}\frac{d\phi(t)}{dt} = (1-\beta) v\Sigma_f \varepsilon \phi(t) pl_f + p\varepsilon l_f \sum_{i=1}^{N} \lambda_i C_i(t) + (DB^2 + \Sigma_a)\phi(t) \tag{9.16}$$

and

$$\frac{dC_i(t)}{dt} = w_i \Sigma_f \phi(t) - \lambda_i C_i(t), \quad i = 1,2,3,\ldots,N \tag{9.17}$$

These reactor kinetic equations are coupled linear first-order ordinary differential equations.

Bibliography

Duderstadt, J. J. and Hamilton, L. J. 1976. *Nuclear Reactor Analysis*. John Wiley & Sons, New York.
Ghoshal, S. N. 2010. *Nuclear Physics*. S. Chand & Co. Ltd, New Delhi, India.
Glasstone, S. and Sesonke, A. 2004. *Nuclear Reactor Engineering*, 4th edn., Vol. 1. CBS Publishers and Distributors Private Limited, New Delhi, India.

Hetrick, D. L. 1971. *Dynamics of Nuclear Reactors*. University of Chicago Press, USA.

Lamarsh, J. R. 1983. *Introduction to Nuclear Engineering*. Addison-Wesley Publishing Company, U.K.

Nayak, S. and Chakraverty, S. 2016. Numerical solution of stochastic point kinetic neutron diffusion equation with fuzzy parameters. *Nuclear Technology* 193(3):444–456.

Sharma, Y. R., Nanda, R. N. and Das, A. K. 2008. *A Textbook of Modern Chemistry*. Kalyani Publisher, New Delhi, India.

Stacey, W. M. 2007. *Nuclear Reactor Physics*. Wiley-VCH, Weinhein, Germany.

10

Stochastic Point Kinetic Diffusion

Let us consider a random experiment with a sample space S. A random variable $X(x)$ is defined as a single-valued real function that assigns a real number called the value of $X(x)$ to each sample point x of S. Often, we use a single letter X for this function in place of $X(x)$.

The sample space S is termed as the domain of the random variable X, and the collection of all the numbers (i.e. values of $X(x)$) is termed as the range of the random variable X. Thus, the range of X is a certain subset of all the real numbers. It may be noted that two or more different sample points might give the same value of $X(x)$. But two different numbers in the range cannot be assigned to the same sample point.

The collection of random variables, which represents the evolution of the system of random values over time, is called stochastic process (Sauer 2012, Ogura 2008, Richardson 2009). In other words, in a probability space, a stochastic process X is a collection $\{X_t \mid t \in T\}$, where X_t is a random variable on the sample space.

10.1 Birth–Death Processes

A birth–death process is a Markov process (a random process whose future probabilities are determined by its most recent values) with the following properties:

1. It is a discrete state space.
2. The states of which can be enumerated with index $i = 0, 1, 2, \ldots$
3. The state transitions can occur only between neighbouring states, $i \to i+1$ or $i \to i-1$.

Consider λ_i and μ_i, where $i = 1, 2, \ldots, n$ be the birth and death rates, respectively. Then, the transition rates are defined in the following way (Figure 10.1):

$$q_{i,j} = \begin{cases} \lambda_i, & j = i+1, \\ \mu_i, & j = i-1, \\ 0, & \text{otherwise.} \end{cases}$$

Let λ_k and μ_k be the birth and death rates in state k, respectively. Then, the probabilities of birth and death in the interval Δt are $\lambda_k \Delta t$ and $\mu_k \Delta t$ for the system in state k, respectively:

1. $P\{\text{state } k \text{ to state } k+1 \text{ in time } \Delta t\} = \lambda_k(\Delta t)$
2. $P\{\text{state } k \text{ to state } k-1 \text{ in time } \Delta t\} = \mu_k(\Delta t)$
3. $P\{\text{state } k \text{ to state } k \text{ in time } \Delta t\} = 1 - (\lambda_k + \mu_k)(\Delta t)$
4. $P\{\text{other transitions in } \Delta t\} = 0$

We have the system state $X(t)$ at time $t = [\text{total births} - \text{total deaths}]$ in $(0, t)$ assuming system states from state 0 at $t = 0$.

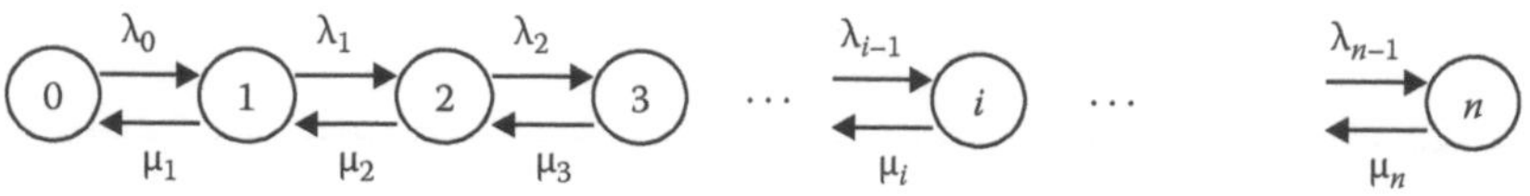

FIGURE 10.1
Transition of states.

Hence,

$$P_0(t + \Delta t) = P_0(t)[1 - \lambda_0 \Delta t] + P_1(t)\mu_1 \Delta t;$$

$$P_k(t + \Delta t) = P_k(t)\left[1 - (\lambda_k + \mu_k)\Delta t\right] + P_{k-1}(t)\lambda_{k-1}\Delta t + P_{k+1}(t)\mu_{k+1}\Delta t.$$

10.2 Pure Death Process

Here, all individuals have the same mortality (death) rate μ, and the transition rates are

$$q_i = \begin{cases} \lambda_i = 0, \\ \mu_i = i\mu, \quad i = 0, 1, 2, \ldots \end{cases}$$

State 0 is an absorbing state, whereas other states are transient, which is shown in Figure 10.2.

10.3 Pure Birth Process

Here, the birth probability per unit time is λ, and the transition rates are

$$q_i = \begin{cases} \lambda_i = \lambda, \\ \mu_i = 0, \quad i = 0, 1, 2, \ldots \end{cases}$$

Initially, the population size is 0 and all the transition rates are shown in Figure 10.3.

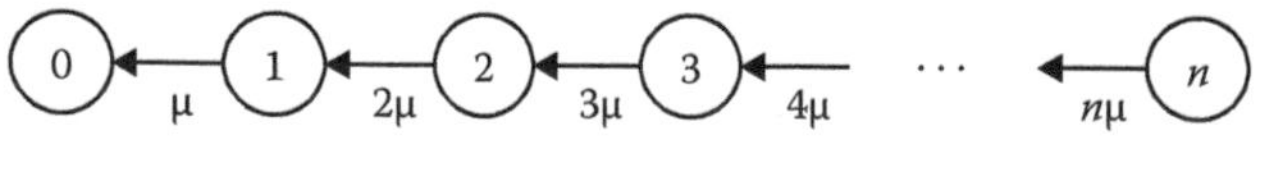

FIGURE 10.2
Transition of states in the pure death process.

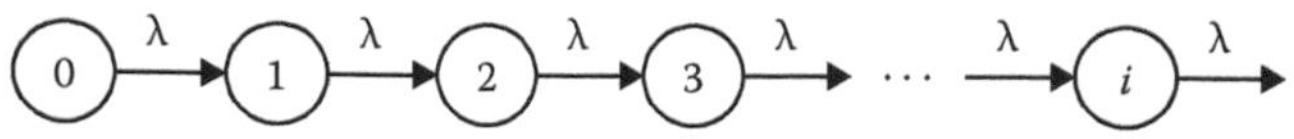

FIGURE 10.3
Transition of states in the pure birth process.

10.4 Stochastic Point Kinetic Model

The point kinetic equation has been modelled in terms of stochastic (Hayes and Allen 2005) by considering the birth and death processes of the neutron and the precursor population. The coupled deterministic time-dependent equations for the neutron density and the delayed neutron precursors may be represented as

$$\frac{\partial N}{\partial t} = Dv\nabla^2 N - \left(\Sigma_a - \Sigma_f\right)vN + \left[(1-\beta)k_\infty \Sigma_a - \Sigma_f\right]vN + \sum_i \lambda_i C_i + S_0 \tag{10.1}$$

$$\frac{\partial C_i}{\partial t} = \beta_i k_\infty \Sigma_a vN - \lambda_i C_i \tag{10.2}$$

for $i = 1, 2, \ldots, m$

where
  $N = N(r,t)$ is the neutron density at position r at time t
  v is the velocity of the neutron
  Σ_f is the neutron fission cross section
  D is the diffusion coefficient
  Σ_a is the absorption coefficient
  $\beta = \sum_{i=1}^{m} \beta_i$ is the delayed neutron fraction
  $1-\beta$ is the prompt neutron fraction
  k_∞ is the infinite medium neutron reproduction factor
  λ_i is the delay constant, $C_i = C_i(r,t)$ density of the ith type of precursors at position r at time t
  S_0 is the extraneous neutron source
  $Dv\nabla^2 N$ is the diffusion term of the neutrons
  $(\Sigma_a - \Sigma_f)$ is the capture cross section
  $[(1-\beta)k_\infty\Sigma_a - \Sigma_f]vN$ is the prompt neutron contribution to the source
  $\sum_{i=1}^{m} \lambda_i C_i$ is the rate of transformations from the neutron precursors to the neutron population

In the present investigation, the neutron captures and the fission process are considered as deaths and pure birth process, respectively. Here, $v(1-\beta)-1$ neutrons are born in each fission along with the precursor fraction $v\beta$.

To apply the separation of variables technique, let us consider $N = f(r)n(t)$ and $C_i = g(r)c_i(t)$, where we assume that N and C_i are separable in time and space. Now, Equation 10.2 becomes

$$\frac{dc_i}{dt} = \beta_i k_\infty \Sigma_a v \frac{f(r)n(t)}{g_i(r)} - \lambda_i c_i(t). \tag{10.3}$$

It is assumed here that f/g_i is independent of time and $(f(r)/g_i(r)) = 1$.

As such, we have

$$\frac{dc_i}{dt} = \beta_i k_\infty \Sigma_a vn - \lambda_i c_i (t).$$

(10.4)

By making the same substitutions as made earlier, Equation 10.1 becomes

$$\frac{dn}{dt} = Dv\frac{\nabla^2 f}{f} n(t) - (\Sigma_a - \Sigma_f)vn(t) + \left[(1-\beta)k_\infty\Sigma_a - \Sigma_f\right]vn(t) + \sum_i \lambda_i \frac{g_i c_i}{f} + \frac{S_0}{f}.$$

(10.5)

We assume that f satisfies $\nabla^2 f + B^2 f = 0$ (a Helmholtz equation) and that S_0 has the same spatial dependence as f. Thus, we will have $q(t) = (S_0(r,t)/f(r))$. Equation 10.5 that describes the rate of change of neutrons with time is

$$\frac{dn}{dt} = -DvB^2 n - (\Sigma_a - \Sigma_f)vn + \left[(1-\beta)k_\infty\Sigma_a - \Sigma_f\right]vn + \sum_i \lambda_i c_i + q.$$

(10.6)

These equations represent a population process where $n(t)$ is the population of neutrons and $c_i(t)$ is the population of the ith precursor. We separate the neutron reactions into two terms: deaths and births. Therefore, we have Equations 10.6 and 10.4 as

$$\frac{dn}{dt} = \underbrace{-DvB^2 n - (\Sigma_a - \Sigma_f)vn}_{\text{Deaths}} + \underbrace{(k_\infty\Sigma_a - \Sigma_f)vn}_{\text{Births}} \underbrace{- \beta k_\infty \Sigma_a vn + \sum_i \lambda_i c_i}_{\text{Transformations}} + q$$

(10.7)

$$\frac{dc_i}{dt} = \beta_i k_\infty \Sigma_a vn - \lambda_i c_i.$$

(10.8)

This system is now to be solved by introducing the following symbols:
The absorption lifetime and the diffusion length are represented as $l_\infty = 1/v\Sigma_a$ and $L^2 = D/\Sigma_a$. Equation 10.7 becomes

$$\frac{dn}{dt} = \frac{-L^2 B^2}{l_\infty} n - \frac{(\Sigma_a - \Sigma_f)}{\Sigma_a l_\infty} n + \frac{(k_\infty\Sigma_a - \Sigma_f)}{\Sigma_a l_\infty} n - \frac{\beta k_\infty}{l_\infty} n + \sum_i \lambda_i c_i + q.$$

(10.9)

After simplification and regrouping, Equation 10.9 becomes

$$\frac{dn}{dt} = \underbrace{\left[\frac{-L^2 B^2 - ((\Sigma_a - \Sigma_f)/\Sigma_a)}{l_\infty}\right]}_{\text{Deaths}} n + \underbrace{\left[\frac{k_\infty - (\Sigma_f/\Sigma_a)}{l_\infty}\right]}_{\text{Births}} n \underbrace{- \frac{\beta k_\infty}{l_\infty} n + \sum_i \lambda_i c_i}_{\text{Transformations}} + q$$

(10.10)

Performing the same substitutions in Equation 10.8 gives

$$\frac{dc_i}{dt} = \frac{\beta_i k_\infty}{l_\infty} n - \lambda_i c_i.$$

(10.11)

Again, two more constants $k = k_\infty / (1 + L^2 B^2)$ and $l_0 = l_\infty / (1 + L^2 B^2)$ are introduced, viz. the reproduction factor and the neutron lifetime, respectively. Now Equation 10.10 becomes

$$\frac{dn}{dt} = \left[\frac{(-1 - L^2 B^2) + (\Sigma_f / \Sigma_a)}{l_\infty} \right] n + \left[\frac{k_\infty}{l_\infty} - \frac{\Sigma_f}{\Sigma_a l_\infty} \right] n - \frac{\beta k_\infty}{l_\infty} n + \sum_i \lambda_i c_i + q. \tag{10.12}$$

After the substitution and rearranging of Equation 10.12, we have

$$\frac{dn}{dt} = \left[\frac{-1}{l_0} + \frac{\Sigma_f}{\Sigma_a l_\infty} \right] n + \left[\frac{k}{l_0} - \frac{\Sigma_f}{\Sigma_a l_\infty} \right] n - \frac{\beta k}{l_0} n + \sum_i \lambda_i c_i + q. \tag{10.13}$$

Similarly, Equation 10.11 will be

$$\frac{dc_i}{dt} = \frac{\beta_i k}{l_0} n - \lambda_i c_i. \tag{10.14}$$

Next, we consider these equations in terms of the neutron generation time. We define $l = l_0 / k$ as the generation time and then Equations 10.13 and 10.14 become

$$\frac{dn}{dt} = \left[\frac{(-1/k)}{l} + \frac{\Sigma_f}{\Sigma_a l_\infty} \right] n + \left[\frac{1}{l} - \frac{\Sigma_f}{\Sigma_a l_\infty} \right] n - \frac{\beta}{l} n + \sum_i \lambda_i c_i + q \tag{10.15}$$

$$\frac{dc_i}{dt} = \frac{\beta_i}{l} n - \lambda_i c_i. \tag{10.16}$$

Next, the reactivity is defined as $\rho = 1 - (1/k)$ and Equation 10.15 becomes

$$\frac{dn}{dt} = \left[\frac{\rho - 1}{l} + \frac{\Sigma_f}{\Sigma_a l_\infty} \right] n + \left[\frac{1}{l} - \frac{\Sigma_f}{\Sigma_a l_\infty} \right] n - \frac{\beta}{l} n + \sum_i \lambda_i c_i + q \tag{10.17}$$

For further simplification, we consider the term $\Sigma_f / (\Sigma_a l_\infty)$, which may be written as

$$\frac{\Sigma_f}{\Sigma_a l_\infty} = \frac{\Sigma_f}{\Sigma_a l_0 (1 + L^2 B^2)} = \frac{\Sigma_f}{\Sigma_a l_0 (k_\infty / k)} = \frac{\Sigma_f}{\Sigma_a l_0 k_\infty} = \frac{1}{l} \frac{\Sigma_f}{\Sigma_a k_\infty} = \frac{\alpha}{l}$$

where α is defined as $\alpha = \Sigma_f / (\Sigma_a k_\infty)$. Equation 10.17 will be

$$\frac{dn}{dt} = \left[\frac{\rho - 1 + \alpha}{l} \right] n + \left[\frac{1 - \alpha - \beta}{l} \right] n + \sum_i \lambda_i c_i + q. \tag{10.18}$$

Finally, this deterministic system may be written as

$$\frac{dn}{dt} = -\left[\frac{-\rho + 1 - \alpha}{l} \right] n + \left[\frac{1 - \alpha - \beta}{l} \right] n + \sum_i \lambda_i c_i + q \tag{10.19}$$

$$\frac{dc_i}{dt} = \frac{\beta_i}{l} n - \lambda_i c_i \tag{10.20}$$

for $i = 1, 2, \dots, m$.

Here, we have considered the population size of neutrons as n and the population size of the ith neutron precursor as c_i. The neutron birth rate due to fission is $b=(1-\alpha-\beta)/(l(-1+(1-\beta)\nu))$, where $(-1+(1-\beta)\nu)$ is the number of new neutrons born in each fission, and $d=(-\rho+1-\alpha)/l$ is the neutron death rate due to captures and leakage, whereas $\lambda_i c_i$ is the rate at which the ith precursor is transformed into neutrons and q is the rate at which the source neutrons are produced.

For simplicity, consider one precursor to derive the stochastic system for the population dynamics. Note that for one precursor $\beta=\beta_1$, β is used for one precursor to represent the total delayed neutron fraction. This notation will be used for generalization. The stochastic system will be extended later for m precursors. The system for one precursor is obtained as

$$\frac{dn}{dt} = -\left[\frac{-\rho+1-\alpha}{l}\right]n + \left[\frac{1-\alpha-\beta}{l}\right]n + \lambda_1 c_1 + q \tag{10.21}$$

$$\frac{dc_1}{dt} = \frac{\beta_1}{l}n - \lambda_1 c_1. \tag{10.22}$$

Now, consider the probability of more than one event occurring during time Δt (a very small time interval). During time Δt, there are four different possibilities for an event for each state defined earlier in the birth and death processes. Let $[\Delta n, \Delta c_1]^T$ be the change in the populations n and c_1 in time Δt, assuming that the changes are approximately normally distributed. Then, the four possibilities for $[\Delta n, \Delta c_1]^T$ may be defined as

$$\begin{bmatrix} \Delta n \\ \Delta c \end{bmatrix}_1 = \begin{bmatrix} -1 \\ 0 \end{bmatrix};$$

$$\begin{bmatrix} \Delta n \\ \Delta c \end{bmatrix}_2 = \begin{bmatrix} -1+(1-\beta)\nu \\ \beta_1\nu \end{bmatrix};$$

$$\begin{bmatrix} \Delta n \\ \Delta c \end{bmatrix}_3 = \begin{bmatrix} 1 \\ -1 \end{bmatrix};$$

$$\begin{bmatrix} \Delta n \\ \Delta c \end{bmatrix}_4 = \begin{bmatrix} 1 \\ 0 \end{bmatrix}.$$

Here, the first event represents a death (capture) of neutrons, the second event is a fission event with $-1+(1-\beta)\nu$ neutrons produced and $\beta_1\nu$ delayed neutrons precursors produced, the third event represents a transformation of a delayed neutron precursor to a neutron and the fourth event is a birth of a source neutron. The probabilities of these four events are

$$P_1 = dn\Delta t;$$

$$P_2 = bn\Delta t = \frac{1}{\nu l}n\Delta t;$$

$$P_3 = \lambda_1 c_1 \Delta t;$$

$$P_4 = q\Delta t;$$

where it is assumed that the extraneous source produces neutrons randomly following a Poisson process with intensity q.

Considering the mean and covariance of the change for a small time interval Δt, we have

$$E\left(\begin{bmatrix} \Delta n \\ \Delta c_1 \end{bmatrix}\right) = \sum_{k=1}^{4} P_k \begin{bmatrix} \Delta n \\ \Delta c_1 \end{bmatrix}_k = \begin{bmatrix} \dfrac{\rho-\beta}{l} n + \lambda_1 c_1 + q \\ \dfrac{\beta_1}{l} n - \lambda_1 c_1 \end{bmatrix} \Delta t. \tag{10.23}$$

Here, we note that

$$(\beta_1 v) P_2 = \frac{\beta_1}{l} n \Delta t \tag{10.24}$$

and

$$b = \frac{1-(1/v)-\beta}{l\left(-1+(1-\beta)v\right)} = \frac{v-1-\beta v}{vl\left(-1+(1-\beta)v\right)} = \frac{1}{vl}$$

assuming $\alpha = 1/v$, we have $b = \alpha/l$.

Finally, we get

$$E\left(\begin{bmatrix} \Delta n \\ \Delta c_1 \end{bmatrix} \begin{bmatrix} \Delta n & \Delta c_1 \end{bmatrix}\right) = \sum_{k=1}^{4} P_k \begin{bmatrix} \Delta n \\ \Delta c_1 \end{bmatrix}_k \begin{bmatrix} \Delta n & \Delta c_1 \end{bmatrix}_k = \hat{B} \Delta t \tag{10.25}$$

where

$$\hat{B} = \begin{bmatrix} \gamma n + \lambda_1 c_1 + q & \dfrac{\beta_1}{l}\left((1-\beta)v-1\right)n - \lambda_1 c_1 \\ \dfrac{\beta_1}{l}\left((1-\beta)v-1\right)n - \lambda_1 c_1 & \dfrac{\beta_1^2 v}{l} n + \lambda_1 c_1 \end{bmatrix},$$

and

$$\gamma = \frac{-1-\rho+2\beta+(1-\beta)^2 v}{l}.$$

Here, the assumption is that the changes are approximately normally distributed and, hence, the result implies

$$\begin{bmatrix} n(t+\Delta t) \\ c_1(t+\Delta t) \end{bmatrix} = \begin{bmatrix} n(t) \\ c_1(t) \end{bmatrix} + \begin{bmatrix} \dfrac{\rho-\beta}{l} n + \lambda_1 c_1 \\ \dfrac{\beta_1}{l} n - \lambda_1 c_1 \end{bmatrix} \Delta t + \begin{bmatrix} q \\ 0 \end{bmatrix} \Delta t + \hat{B}^{\frac{1}{2}} \sqrt{\Delta t} \begin{bmatrix} \eta_1 \\ \eta_2 \end{bmatrix}, \tag{10.26}$$

where $\eta_1, \eta_2 \sim N(0,1)$ and $\hat{B}^{1/2}$ is the square root of the matrix $\hat{B}$, where $\hat{B} = \hat{B}^{1/2} \cdot \hat{B}^{1/2}$.

As $\Delta t \to 0$, we will have the following Ito stochastic system:

$$\frac{d}{dt}\begin{bmatrix} n \\ c_1 \end{bmatrix} = \hat{A}\begin{bmatrix} n \\ c_1 \end{bmatrix} + \begin{bmatrix} q \\ 0 \end{bmatrix} + \hat{B}^{1/2}\frac{d\vec{W}}{dt} \tag{10.27}$$

where

$$\hat{A} = \begin{bmatrix} \dfrac{\rho-\beta}{l} & \lambda_1 \\ \dfrac{\beta_1}{l} & -\lambda_1 \end{bmatrix}, \quad \hat{B} = \begin{bmatrix} \gamma n + \lambda_1 c_1 + q & \dfrac{\beta_1}{l}\big((1-\beta)v-1\big)n - \lambda_1 c_1 \\ \dfrac{\beta_1}{l}\big((1-\beta)v-1\big)n - \lambda_1 c_1 & \dfrac{\beta_1^2 v}{l}n + \lambda_1 c_1 \end{bmatrix},$$

and

$$\vec{W} = \vec{W}(t) = \begin{bmatrix} W_1(t) \\ W_2(t) \end{bmatrix}.$$

Here, $W_1(t)$ and $W_2(t)$ are Wiener processes.

One may generalize this concept into m precursors. Then, the matrices will be

$$\hat{A} = \begin{bmatrix} \dfrac{\rho-\beta}{l} & \lambda_1 & \lambda_2 & \cdots & \lambda_m \\ \dfrac{\beta_1}{l} & -\lambda_1 & 0 & \cdots & 0 \\ \dfrac{\beta_2}{l} & 0 & -\lambda_2 & \ddots & \vdots \\ \vdots & \vdots & \ddots & \ddots & 0 \\ \dfrac{\beta_m}{l} & 0 & \cdots & 0 & -\lambda_m \end{bmatrix} \quad \text{and} \quad \hat{B} = \begin{bmatrix} \zeta & a_1 & a_2 & \cdots & a_m \\ a_1 & r_1 & b_{2,3} & \cdots & b_{2,m} \\ a_2 & 0 & r_2 & \ddots & \vdots \\ \vdots & \vdots & \ddots & \ddots & b_{m-1,m} \\ a_m & b_{m,2} & \cdots & b_{m,m-1} & r_m \end{bmatrix}$$

where

$$\zeta = \gamma n + \sum_{i=1}^{m}\lambda_i c_i + q, \quad a_i = \frac{\beta_i}{l}\big((1-\beta)v-1\big)n - \lambda_i c_i,$$

$$b_{i,j} = \frac{\beta_{i-1}\beta_{j-1}v}{l}n, \quad \text{and} \quad r_i = \frac{\beta_i^2 v}{l}n + \lambda_i c_i.$$

Using a similar approach for m precursors, one may get the following Ito stochastic system:

$$\frac{d}{dt}\begin{bmatrix} n \\ c_1 \\ c_2 \\ \vdots \\ c_m \end{bmatrix} = \hat{A}\begin{bmatrix} n \\ c_1 \\ c_2 \\ \vdots \\ c_m \end{bmatrix} + \begin{bmatrix} q \\ 0 \\ 0 \\ \vdots \\ 0 \end{bmatrix} + \hat{B}^{1/2}\frac{d\vec{W}}{dt}. \tag{10.28}$$

Equation 10.28 is called the stochastic point kinetic equations for m precursors. In Equation 10.28, we may note that if $\hat{B} = 0$, then Equation 10.28 transforms into the standard deterministic point kinetic equations, hence Equation 10.28 can be considered a generalization of the standard point kinetic model.

Bibliography

Hayes, J. G. and Allen, E. J. 2005. Stochastic point kinetic equations in nuclear reactor dynamics. *Annals of Nuclear Energy* 32:572–587.

Higham, D. J. and Kloeden, P. 2005. Numerical methods for nonlinear stochastic differential equations with jumps. *Numerische Mathematik* 101:101–119.

Kloeden, P. and Platen, E. 1992. *Numerical Solution of Stochastic Differential Equations*. Springer, Berlin, Germany.

Laguerre, O. and Flick, D. 2010. Temperature prediction in domestic refrigerators: Deterministic and stochastic approaches. *International Journal of Refrigeration* 33:41–51.

Malinowski, M. T. and Michta, M. 2011. Stochastic fuzzy differential equations with an application. *Kybernetika* 47(1):123–143.

Ogura, Y. 2008. On stochastic differential equations with fuzzy set coefficients. In: Dubois, D., et al. (eds.), *Soft Methods for Handling Variability and Imprecision*. Springer, Berlin, Germany.

Oksendal, B. 2003. *Stochastic Differential Equations: An Introduction with Applications*. Springer-Verlag, Heidelberg, Germany.

Richardson, M. 2009. *Stochastic Differential Equations Case Study*.

Sauer, T. 2012. *Numerical Solution of Stochastic Differential Equations in Finance*. Springer, USA.

Tochastic, S. and Odeling, M. 2007. *Stochastic Computational Fluid Mechanics*.

11

Hybridized Uncertainty in Point Kinetic Diffusion

As mentioned earlier, every system possesses uncertainties and they occur due to partial knowledge and truth. In general, partial knowledge–based uncertainties may be handled by probability theory and truth-based uncertainties are operated through possibility theory. In practical problems, we may have the combined effect of both the uncertainties. As such, this chapter comprises hybridization of the concept of stochasticity with fuzzy theory. Here, we have the modelled SDE, which includes the essence of fuzziness, and we call it the fuzzy stochastic differential equation (FSDE). Two well-known methods, viz. Euler–Maruyama method (EMM) and Milstein method (MM), are extended to the fuzzy form, and these are named fuzzy Euler–Maruyama method (FEMM) and fuzzy Milstein method (FMM). These methods have been used to investigate various diffusion problems.

11.1 Black–Scholes Stochastic Differential Equation

The Black–Scholes SDE with fuzzy uncertainty is investigated here. The uncertainties are assumed to occur due to the parameters involved in the system and these are considered as triangular fuzzy numbers (TFNs). The fuzzy arithmetic (Nayak and Chakraverty 2013) is used as a tool to handle FSDE. In particular, a system of Ito stochastic differential equations is analyzed with the fuzzy parameters. Furthermore, exact and EM approximation methods with fuzzy uncertainties are demonstrated.

For the sake of completeness, initially we discuss the crisp SDE and it is solved by using the Ito integral technique. Furthermore, the same problems are discussed for uncertain cases and the corresponding FSDEs are solved.

11.1.1 Preliminary

Let us consider a standard stochastic differential equation:

$$dX = a(t, X)dt + b(t, X)dW_t \tag{11.1}$$

which is written in differential form.

The integral form of Equation 11.1 becomes

$$X(t) = X(0) + \int_0^t a(s, y)ds + \int_0^t b(s, y)dW_s \tag{11.2}$$

where the last term on the right-hand side of Equation 11.2 is called the Ito integral.

We take $c = t_0 < t_1 < t_2 < \cdots < t_{n-1} < t_n = d$ as a grid of points on an interval $[c, d]$, then the Ito integral may be defined in the following limit form:

$$\int_c^d f(t) dW_t = \lim_{\Delta t \to 0} \sum_{i=1}^n f(t_{i-1}) \Delta W_i \tag{11.3}$$

where $\Delta W_i = W_{t_i} - W_{t_{i-1}}$, a step of Brownian motion across the interval.

11.1.2 Analytical Solution of Stochastic Differential Equations

First, Equation 11.1 is solved analytically by using the Ito formula. If X_t is an Ito process, then we have

$$dX_t = udt + vdW_t \tag{11.4}$$

Let $g(t, x) \in C^2([0, \infty] \times \mathfrak{R})$ (i.e. g is twice continuous differentiable on $[0, \infty] \times \mathfrak{R}$). Then $Y_t = g(t, X_t)$ is again an Ito process (Oksendal 2003) and one can write

$$dY_t = \frac{\partial g}{\partial t}(t, X_t) dt + \frac{\partial g}{\partial x}(t, X_t) dX_t + \frac{1}{2} \frac{\partial^2 g}{\partial x^2}(t, X_t)(dX_t)^2 \tag{11.5}$$

where $(dX_t)^2 = (dX_t) \cdot (dX_t)$, which is computed as follows:

$$dt \cdot dt = dt \cdot dW_t = dW_t \cdot dt = 0;$$
$$dW_t \cdot dW_t = dt.$$

Example 11.1

Let us consider an SDE (Malinowski and Michta 2011):

$$\begin{cases} \dfrac{dX_t}{dt} = a_t X_t, \\ X(0, x) = X_0 \end{cases} \tag{11.6}$$

where
$a_t = r_t + \alpha W_t$
W_t and α are noise and constant, respectively

Equation 11.6 may be written as

$$\frac{dX_t}{dt} = (r_t + \alpha W_t) X_t$$
$$= r_t X_t + \alpha W_t X_t$$
$$\Rightarrow dX_t = r_t X_t dt + \alpha X_t dB_t \quad (\because W_t \cdot dt = dB_t)$$
$$\Rightarrow \frac{dX_t}{X_t} = r_t dt + \alpha dB_t$$
$$\Rightarrow \int_0^t \frac{dX_s}{X_s} = r_t t + \alpha B_t$$

Using the Ito formula for the function $g(t, x) = \ln x$ (Oksendal 2003), we get the following:

$$d\left(\ln X_t\right) = \frac{1}{X_t} dX_t - \frac{1}{2}\left(\frac{1}{X_t^2}\right)\left(dX_t\right)^2$$

$$= \frac{1}{X_t} dX_t - \frac{1}{2}\left(\frac{1}{X_t^2}\right)\alpha^2 X_t^2 dt$$

$$= \frac{1}{X_t} dX_t - \frac{1}{2}\alpha^2 dt$$

Integrating this, we get

$$\int_0^t d\left(\ln X_t\right) = \int \frac{dX_t}{X_t} - \int \frac{1}{2}\alpha^2 dt$$

$$\Rightarrow \ln X_t - \ln X_0 = \int r_t dt + \alpha dB_t - \frac{1}{2}\alpha^2 t$$

$$\Rightarrow \ln \frac{X_t}{X_0} = \left(r_t - \frac{1}{2}\alpha^2\right)t + \alpha B_t$$

$$\Rightarrow X_t = X_0 e^{\left(r_t - \frac{1}{2}\alpha^2\right)t + \alpha B_t}$$

It may be noted from the literature that except for some standard problems, the exact method may not be applicable for others. Hence, we need a numerical treatment to handle non-trivial problems, and this has been discussed in the following sections.

11.1.3 Solution of Fuzzy Stochastic Differential Equations

Let us consider an SDE with fuzzy parameters, then Equation 11.1 may be written as

$$d\left[\underline{X}(\alpha),\ \bar{X}(\alpha)\right] = \left[\underline{a}(\alpha),\ \bar{a}(\alpha)\right]dt + \left[\underline{b}(\alpha),\ \bar{b}(\alpha)\right]dW_t \tag{11.7}$$

Equation 11.7 is solved here by exact as well as numerical methods.

Using the limit method (Chakraverty and Nayak 2013), the FSDE (11.7) may be represented in modified limit form as

$$d\left[\lim_{s\to\infty} X(\alpha),\ \lim_{s\to1} X(\alpha)\right] = \left[\lim_{s\to\infty} a(\alpha),\ \lim_{s\to1} a(\alpha)\right]dt + \left[\lim_{s\to\infty} b(\alpha),\ \lim_{s\to1} b(\alpha)\right]dW_t \tag{11.8}$$

where

$$\tilde{X}(\alpha) = \underline{X}(\alpha) + \frac{\bar{X}(\alpha) - \underline{X}(\alpha)}{s}$$

$$\tilde{a}(\alpha) = \underline{a}(\alpha) + \frac{\bar{a}(\alpha) - \underline{a}(\alpha)}{s}$$

and

$$\tilde{b}(\alpha) = \underline{b}(\alpha) + \frac{\bar{b}(\alpha) - \underline{b}(\alpha)}{s}$$

Initially, for the exact (or crisp) case, we take the crisp representation of $\tilde{X}(\alpha)$, $\tilde{a}(\alpha)$, $\tilde{b}(\alpha)$ and use the Ito integral to solve the problem.

Now, if we apply the fuzzy concept for the EMM, then we get

$$\tilde{X}_0(\alpha) = \tilde{w}_0(\alpha)$$
$$\tilde{w}_{i+1}(\alpha) = \tilde{w}_i(\alpha) + \tilde{a}(t_i, w_i, \alpha)\Delta t_{i+1} + \tilde{b}(t_i, w_i, \alpha)\Delta W_i \tag{11.9}$$

where

$$\tilde{X}_0(\alpha) = \underline{X_0}(\alpha) + \frac{\overline{X}_0(\alpha) - \underline{X_0}(\alpha)}{s}$$

$$\tilde{w}_0(\alpha) = \underline{w_0}(\alpha) + \frac{\overline{w}_0(\alpha) - \underline{w_0}(\alpha)}{s}$$

$$\tilde{w}_{i+1}(\alpha) = \underline{w_{i+1}}(\alpha) + \frac{\overline{w}_{i+1}(\alpha) - \underline{w_{i+1}}(\alpha)}{s}$$

$$\tilde{a}(t_i, w_i, \alpha) = \underline{a}(t_i, w_i, \alpha) + \frac{\overline{a}(t_i, w_i, \alpha) - \underline{a}(t_i, w_i, \alpha)}{s}$$

$$\tilde{b}(t_i, w_i, \alpha) = \underline{b}(t_i, w_i, \alpha) + \frac{\overline{b}(t_i, w_i, \alpha) - \underline{b}(t_i, w_i, \alpha)}{s}$$

By applying $\lim_{s \to \infty}$ and $\lim_{s \to 1}$ on the solution, we get the left and right bounds, whereas we obtain various solution sets by considering different values of $\alpha \in [0, 1]$. It is noticed that sometimes we get weak solutions (the left- and right-bound solutions overlap or intersect each other), and this occurs due to the randomness of the system. This may easily be observed from the following example problems.

11.1.4 Example Problems

In this section, we have considered two example problems and taken the parameters as fuzzy. Initially, the problem is studied for crisp parameters for both the exact and numerical methods and then the fuzzy parameters are incorporated.

Example 11.2

Consider the Black–Scholes SDE (Black and Scholes 1973).
The crisp EM approximation for this SDE is as follows:

$$w_0 = X_0$$
$$w_{i+1} = w_i + \mu w_i \Delta t_i + \sigma w_i \Delta W_i \tag{11.10}$$

where the values of the involved parameters are given in Table 11.1.

TABLE 11.1

Crisp and Fuzzy Values of the Involved Parameters

Parameters	Crisp	TFN
μ	0.75	[0.65, 0.75, 0.85]
σ	0.30	[0.25, 0.30, 0.35]

Initially, the Black–Scholes SDE has been solved for the crisp parameter and then the fuzzy parameters are considered for investigation. Here, we compute a discretized Brownian path over [0, 1] with $\delta t = 2^{-8}$ and the obtained solution is plotted with a solid light gray line in Figure 11.1. We then apply the EMM using a step size $\Delta t = R\delta t$, where R is a constant (with $R = 4$) and the solution is presented in Figure 11.1 with a dark gray line.

Now, drift (μ) and diffusion (σ) coefficients are taken as TFN, which are given in Table 11.1. The exact method (Ito process) is used to obtain the solution that is depicted in Figure 11.2. Here, the black and light gray solid lines represent the left and right bounds of the uncertainties. Next, the left and right values of the uncertainties are plotted with the exact solutions in Figures 11.3 and 11.4, respectively. Then the EMM is used to solve the uncertain SDE and the results are graphically depicted in Figure 11.5, where the black and light gray lines represent the left and right bounds of the uncertain solutions. The region covered in between the left and right bounds is the uncertain solution of the Black–Scholes SDE. In Figure 11.6, we have given the left and right bounds, which represent the uncertain solution of the Black–Scholes SDE along with the crisp solution, and we found that the exact solution lies within the region covered by the left and right solutions. Furthermore, the EMM is used for the next example (Figure 11.7).

Example 11.3

The SDE of the Langevin equation is

$$dX(t) = -\mu X(t)dt + \sigma dW_t \qquad (11.11)$$

where μ and σ are positive constants.

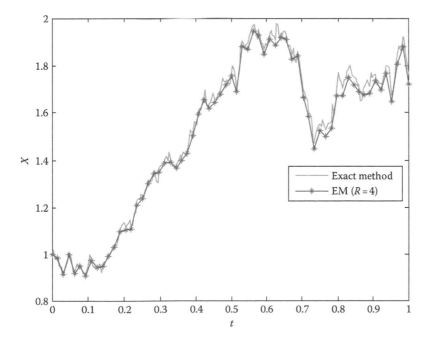

FIGURE 11.1
Solution of the Black–Scholes SDE when parameters are crisp.

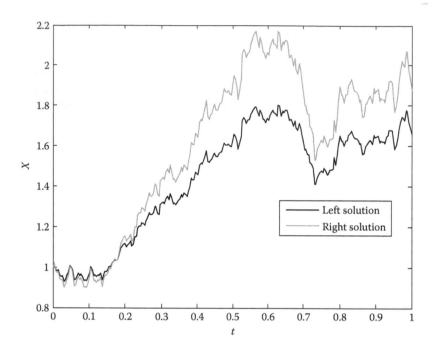

FIGURE 11.2
Ito solution of the Black–Scholes SDE when parameters are fuzzy.

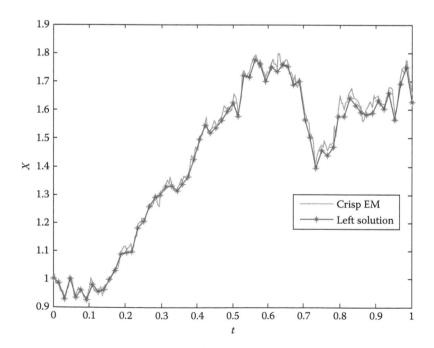

FIGURE 11.3
Crisp EM solution of the Black–Scholes SDE and the left uncertain bound solutions.

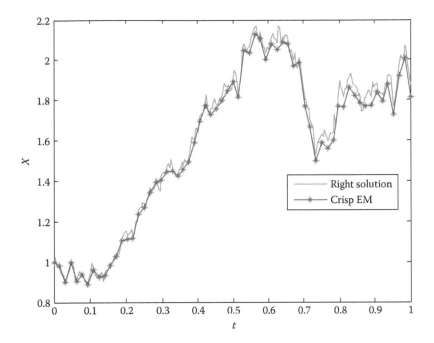

FIGURE 11.4
Crisp EM solution of the Black–Scholes SDE and the right uncertain bound solutions.

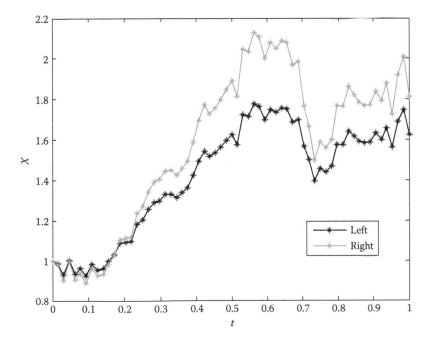

FIGURE 11.5
EM solution of the Black–Scholes SDE when parameters are fuzzy.

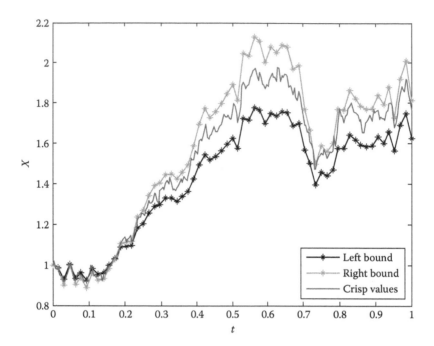

FIGURE 11.6
EM solution of the Black–Scholes SDE when parameters are fuzzy with the Ito solution.

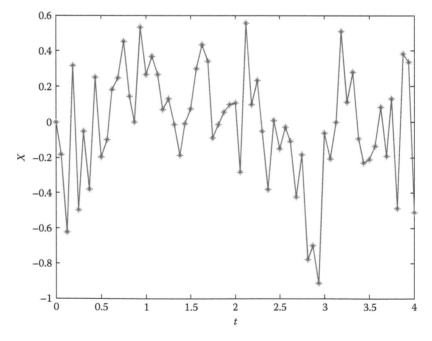

FIGURE 11.7
EM solution of the Langevin SDE when parameters are fuzzy.

The EM approximation for Equation 11.11 is

$$w_0 = X_0$$
$$w_{i+1} = w_i - \mu w_i \Delta t_i + \sigma \Delta W_i \tag{11.12}$$

The same parameter values that are used in Equation 11.8 are considered here and those are given in Table 11.2.

In Figure 11.8, we have given a plot for the solution of the Langevin SDE when parameters are crisp, whereas the solution for the Langevin SDE is presented in Figure 11.9, where the parameters are taken as fuzzy. The left- and right-bound solutions are shown in blue and magenta colours, respectively.

For better visualization of uncertain distribution of EM approximation results, fuzzy plots are represented in Figures 11.9 and 11.10 for Examples 11.2 and 11.3, respectively.

One may also see that the uncertain widths are randomly varying. It may be noted that if the uncertainty of the parameter changes, the uncertain width of the solution sets varies accordingly.

TABLE 11.2

Crisp and Fuzzy Values of the Used Parameters (for Section 11.1.4)

Parameters	Crisp	TFN
μ	10	[8, 10, 12]
σ	1	[0.5, 1, 1.5]

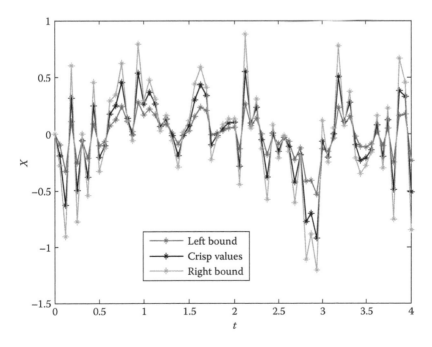

FIGURE 11.8
EM solution of the Langevin SDE when parameters are fuzzy with the crisp solution.

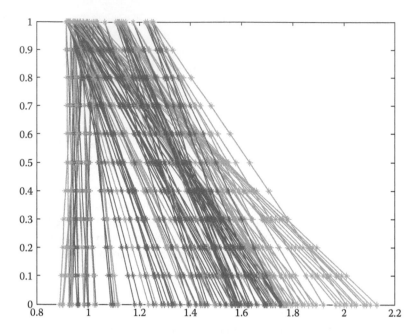

FIGURE 11.9
Fuzzy plot of EM solution of the Black–Scholes SDE when parameters are TFN (Example 11.2).

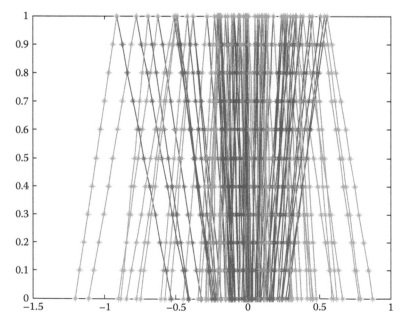

FIGURE 11.10
Fuzzy plot of EM solution of the Langevin SDE when parameters are TFN (Example 11.3).

11.2 Langevin Stochastic Differential Equation

Here, two different approaches, viz. FEMM and FMM, are presented for solving the uncertain Langevin SDE. The uncertainties are taken in their initial conditions as well as associated parameters in terms of TFN. The limit method for fuzzy arithmetic has been used as a tool to handle the FSDE.

11.2.1 Solution of Fuzzy Stochastic Differential Equations

As mentioned earlier that the uncertainties are taken in initial conditions as well as associated parameters, the problems may be classified into the following three cases.

Case 1

For the first case, we have considered the initial conditions of the SDE as uncertain, viz. fuzzy. As such, we consider Equation 11.8 with initial condition as fuzzy and then we have

$$\begin{cases} dX(t) = a(t, X)dt + b(t, X)dW_t \\ \tilde{X}(0) = \left[\underline{X_0}(\alpha),\ \overline{X}_0(\alpha) \right] \end{cases} \tag{11.13}$$

Now, if we apply this fuzzy concept for the EMM, then Equation 11.13 may be represented in the following way:

$$\tilde{X}_0(\alpha) = \tilde{w}_0(\alpha) \tag{11.14}$$
$$\tilde{w}_{i+1}(\alpha) = \tilde{w}_i(\alpha) + \tilde{a}(t_i, w_i)\Delta t_{i+1} + \tilde{b}(t_i, w_i)\Delta W_i$$

where

$$\tilde{X}_0(\alpha) = \underline{X_0}(\alpha) + \frac{\overline{X}_0(\alpha) - \underline{X_0}(\alpha)}{s}$$

$$\tilde{w}_0(\alpha) = \underline{w_0}(\alpha) + \frac{\overline{w}_0(\alpha) - \underline{w_0}(\alpha)}{s}$$

Furthermore, it is noticed that due to the fuzziness of the initial value, we get a series of approximated fuzzy solutions and these are

$$\tilde{w}_{i+1}(\alpha) = \underline{w_{i+1}}(\alpha) + \frac{\overline{w}_{i+1}(\alpha) - \underline{w_{i+1}}(\alpha)}{s}, \quad i \geq 0.$$

When the MM is used to handle this situation (Equation 11.13), then the approximation scheme may be represented in the following manner:

$$\tilde{X}_0(\alpha) = \tilde{w}_0(\alpha) \tag{11.15}$$
$$\tilde{w}_{i+1}(\alpha) = \tilde{w}_i(\alpha) + \tilde{a}(t_i, w_i)\Delta t_{i+1} + \tilde{b}(t_i, w_i)\Delta W_i + \frac{1}{2}\tilde{b}(t_i, w_i)\frac{\partial \tilde{b}}{\partial x}(t_i, w_i)\left(\Delta W_i^2 - \Delta t_i\right)$$

Case 2

In this case, we assume the involve parameters (or the coefficients) only as fuzzy, then the modified form of Equation 11.8 may be written as

$$\begin{cases} d\tilde{X}(t) = \left[\underline{a}(t, X, \alpha), \ \overline{a}(t, X, \alpha)\right]dt + \left[\underline{b}(t, X, \alpha), \ \overline{b}(t, X, \alpha)\right]dW_t \\ X(0) = X_0 \end{cases} \tag{11.16}$$

By applying the fuzzy concept for the EMM, Equation 11.16 may be represented as

$$\begin{aligned} X_0 &= w_0 \\ \tilde{w}_{i+1} &= \tilde{w}_i + \tilde{a}(t_i, w_i, \ \alpha)\Delta t_{i+1} + \tilde{b}(t_i, w_i, \ \alpha)\Delta W_i \end{aligned} \tag{11.17}$$

where

$$\tilde{a}(t_i, w_i, \alpha) = \underline{a}(t_i, w_i, \alpha) + \frac{\overline{a}(t_i, w_i, \alpha) - \underline{a}(t_i, w_i, \alpha)}{s}$$

$$\tilde{b}(t_i, w_i, \alpha) = \underline{b}(t_i, w_i, \alpha) + \frac{\overline{b}(t_i, w_i, \alpha) - \underline{b}(t_i, w_i, \alpha)}{s}$$

For the MM, the approximation scheme for Equation 11.16 may be written as

$$\begin{aligned} X_0 &= w_0 \\ \tilde{w}_{i+1} &= \tilde{w}_i + \tilde{a}(t_i, w_i, \ \alpha)\Delta t_{i+1} + \tilde{b}(t_i, \ w_i, \ \alpha)\Delta W_i + \frac{1}{2}\tilde{b}(t_i, \ w_i, \ \alpha)\frac{\partial b}{\partial x}(t_i, \ w_i, \ \alpha)\left(\Delta W_i^2 - \Delta t_i\right) \end{aligned} \tag{11.18}$$

It may be pointed out that as the parameters are taken as fuzzy, we get a series of approximated fuzzy solutions in the α-cut form and these are $\tilde{w}_{i+1}(\alpha) = \underline{w}_{i+1}(\alpha) + \dfrac{\overline{w}_{i+1}(\alpha) - \underline{w}_{i+1}(\alpha)}{s}$.

Case 3

Finally, both the initial condition and involved parameters are considered as fuzzy and then Equation 11.8 may be represented as

$$\begin{cases} d\tilde{X}(t) = \left[\underline{a}(t, X, \alpha), \ \overline{a}(t, X, \alpha)\right]dt + \left[\underline{b}(t, X, \alpha), \ \overline{b}(t, X, \alpha)\right]dW_t \\ \tilde{X}(0) = \left[\underline{X}_0(\alpha), \overline{X}_0(\alpha)\right] \end{cases} \tag{11.19}$$

Applying the same fuzzy concept for the EMM, Equation 11.19 may now be written as

$$\begin{aligned} \tilde{X}_0(\alpha) &= \tilde{w}_0(\alpha) \\ \tilde{w}_{i+1}(\alpha) &= \tilde{w}_i(\alpha) + \tilde{a}(t_i, w_i, \ \alpha)\Delta t_{i+1} + \tilde{b}(t_i, w_i, \ \alpha)\Delta W_i \end{aligned} \tag{11.20}$$

and for the MM, the approximation scheme becomes

$$\tilde{X}_0(\alpha) = \tilde{w}_0(\alpha)$$

$$\tilde{w}_{i+1}(\alpha) = \tilde{w}_i(\alpha) + \tilde{a}(t_i, w_i, \alpha)\Delta t_{i+1} + \tilde{b}(t_i, w_i, \alpha)\Delta W_i$$

$$+ \frac{1}{2}\tilde{b}(t_i, w_i, \alpha)\frac{\partial b}{\partial x}(t_i, w_i, \alpha)\left(\Delta W_i^2 - \Delta t_i\right) \qquad (11.21)$$

11.2.2 Example Problem

Let us consider the SDE of the Langevin equation

$$dX(t) = -\mu X(t)dt + \sigma X dW_t \qquad (11.22)$$

where μ and σ are positive constants.

The parameters for Equation 11.22 in terms of crisp and fuzzy are given in Table 11.3.

Case 1

In this case, only the initial condition is fuzzy. As mentioned earlier, EM and MM are used to handle the problem. The investigated results are depicted in Figures 11.11 through 11.14. FEMM solutions are given in Figures 11.11 and 11.12 for $\alpha = 0.5$ and 0 (interval). Similarly, Figures 11.13 and 11.14 depict the FMM results for $\alpha = 0.5$ and 0 (interval).

Case 2

Here, the involved parameters (except the initial condition) are considered only as fuzzy and the solutions for the FEMM and the FMM are shown in Figures 11.15 through 11.18. The values of alpha (α) are 0.5 and 0 as mentioned in Case 1.

Case 3

Finally, both the initial condition and parameters are taken as fuzzy and the problem is investigated using the FEMM and the MM. The solutions can be visualized from Figures 11.19 through 11.22 for the same values of α as in the earlier cases.

The obtained results in digital form are also incorporated in Table 11.4. The left, centre and right values of the TFNs are given for the different cases. Furthermore, the values of X at different time ($t = 1, 2, 3, 4$) denoted as $X(1)$, $X(2)$, $X(3)$ and $X(4)$ are presented in this table for both the FEMM and the FMM.

TABLE 11.3

Crisp and Fuzzy Values of the Used Parameters (for Section 11.2.2)

Parameters	Crisp	TFN
X_0	1	[0.5, 1, 1.5]
μ	15	[10, 15, 20]
σ	1	[0.5, 1, 1.5]

TABLE 11.4

Fuzzy Solution of the Problem for Different Cases

		Crisp		Left		Centre		Right	
		EMM	MM	FEMM	FMM	FEMM	FMM	FEMM	FMM
Case 1	X(1)	0.1003	0.0847	0.1337	0.0847	0.1003	0.0847	0.1003	0.0847
	X(2)	0.0732	0.0708	0.0523	0.0708	0.0732	0.0708	0.0732	0.0708
	X(3)	0.2367	0.2250	−0.0319	0.2250	0.2367	0.2250	0.2367	0.2250
	X(4)	−0.6246	−0.5883	−0.2565	−0.5883	−0.6246	−0.5883	−0.6246	−0.5883
Case 2	X(1)	0.1003	0.0847	0.1337	0.1316	0.1003	0.0847	−0.1029	−0.1525
	X(2)	0.0732	0.0708	0.0523	0.0519	0.0732	0.0708	0.0770	0.0700
	X(3)	0.2367	0.2250	−0.0319	−0.0399	0.2367	0.2250	0.6432	0.6647
	X(4)	−0.6246	−0.5883	−0.2565	−0.2507	−0.6246	−0.5883	−0.9620	−0.9499
Case 3	X(1)	0.1003	0.0847	0.1003	0.1316	0.1003	0.0847	−0.1029	−0.1525
	X(2)	0.0732	0.0708	0.0732	0.0519	0.0732	0.0708	0.0770	0.0700
	X(3)	0.2367	0.2250	0.2367	−0.0399	0.2367	0.2250	0.6432	0.6647
	X(4)	−0.6246	−0.5883	−0.6246	−0.2507	−0.6246	−0.5883	−0.9620	−0.9499

The sensitiveness of the solution set is studied in terms of the uncertain widths. Here, we have considered the fuzzy solutions and widths of X at different time ($t = 1, 2, 3, 4$), which are encrypted in Table 11.5. The uncertain widths for both FEMM and FMM are given in Table 11.5, and it may be observed that the FMM gives less width.

Next, the uncertain solutions (X at $t = 2$) are plotted in terms of TFN for various cases in Figures 11.23 through 11.25.

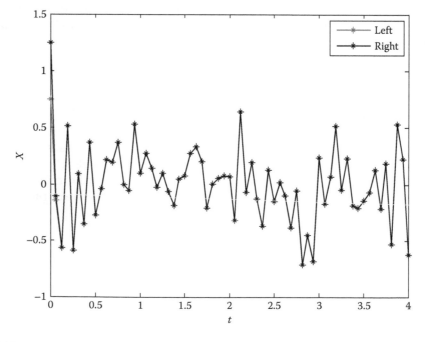

FIGURE 11.11

Interval solutions using the FEMM at $\alpha = 0.5$.

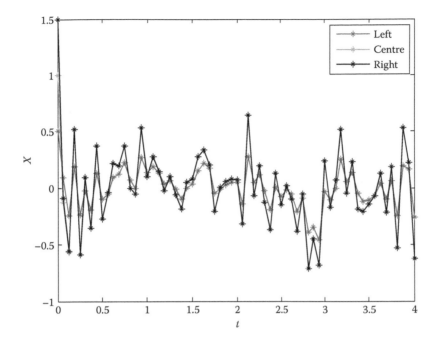

FIGURE 11.12
Fuzzy solutions using the FEMM.

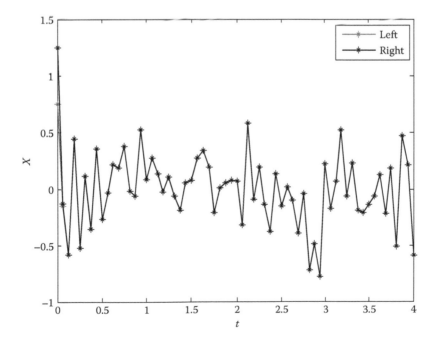

FIGURE 11.13
Interval solutions using the FMM at α = 0.5.

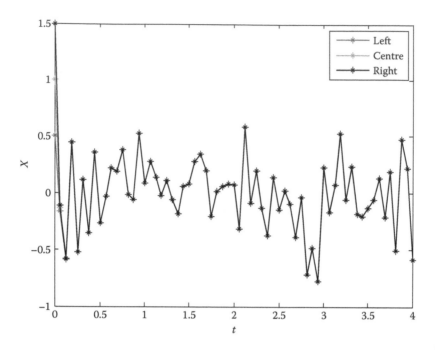

FIGURE 11.14
Fuzzy solutions using the FMM.

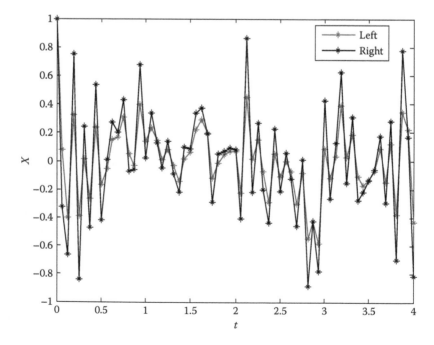

FIGURE 11.15
Interval solutions using the FEMM at α = 0.5.

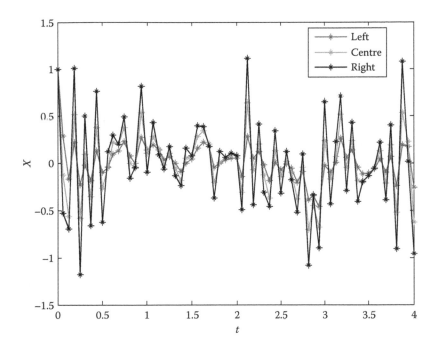

FIGURE 11.16
Fuzzy solutions using the FEMM.

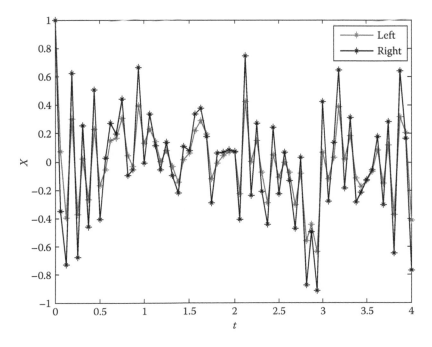

FIGURE 11.17
Interval solutions using the FMM at $\alpha = 0.5$.

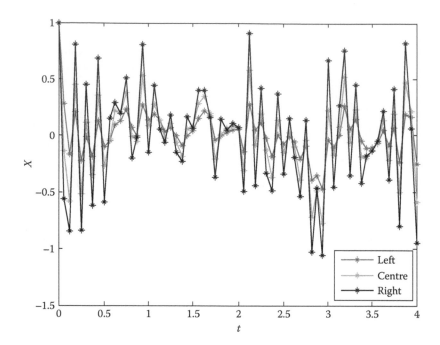

FIGURE 11.18
Fuzzy solutions using the FMM.

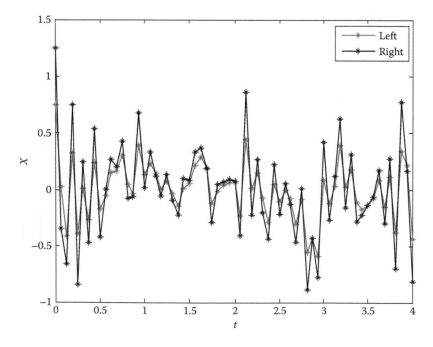

FIGURE 11.19
Interval solutions using the FEMM at $\alpha = 0.5$.

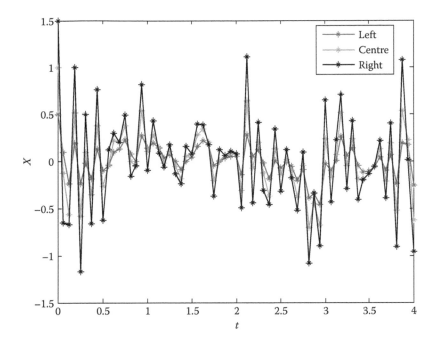

FIGURE 11.20
Fuzzy solutions using the FEMM.

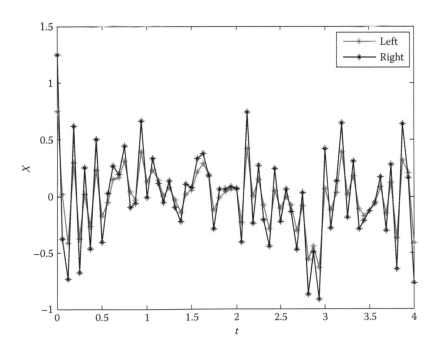

FIGURE 11.21
Interval solutions using the FMM at $\alpha = 0.5$.

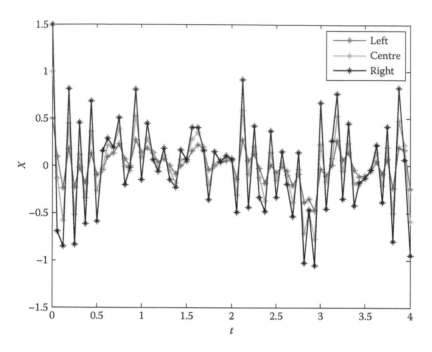

FIGURE 11.22
Fuzzy solutions using the FMM.

In Figure 11.23, it is seen that the solution obtained by using the FMM gives a TFN (where the left and right solutions are same), which is parallel to the membership function axis. Again, it has also been seen that the non-increasing left continuous function of the obtained TFN by using the FFEM becomes parallel to the membership function axis, whereas in Figure 11.24, the left non-decreasing function values of the resultant TFNs are more approximate as compared to the left non-increasing function values.

TABLE 11.5

Width of the Solutions at $\alpha = 0$ Using the FEMM and FMM

		Width	
		FEMM	**FMM**
Case 1	$X(1)$	0.0334	0
	$X(2)$	0.0209	0
	$X(3)$	0.2686	0
	$X(4)$	0.3681	0
Case 2	$X(1)$	0.2366	0.2841
	$X(2)$	0.0247	0.0189
	$X(3)$	0.6751	0.7046
	$X(4)$	0.7055	0.6992
Case 3	$X(1)$	0.2032	0.2841
	$X(2)$	0.0038	0.0189
	$X(3)$	0.4065	0.7046
	$X(4)$	0.3374	0.6992

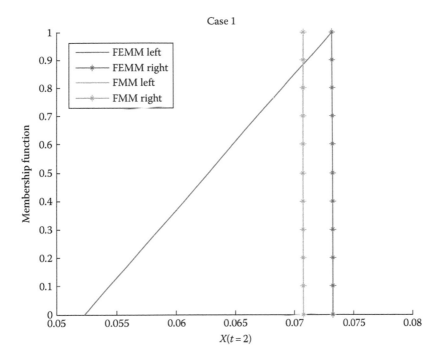

FIGURE 11.23
Fuzzy plot for Case 1 at $X(t = 2)$.

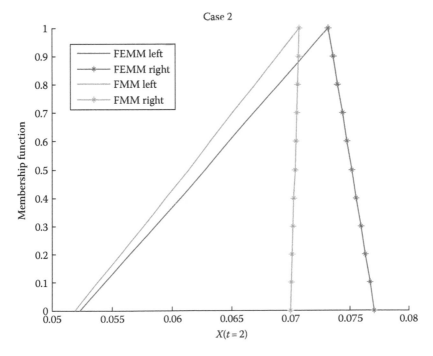

FIGURE 11.24
Fuzzy plot for Case 2 at $X(t = 2)$.

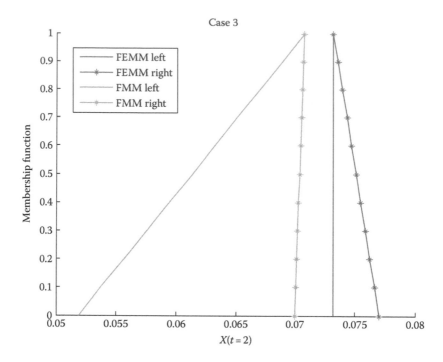

FIGURE 11.25
Fuzzy plot for Case 3 at $X(t = 2)$.

For Case 3, Figure 11.25 gives the straight line for the non-increasing and non-decreasing left continuous function using FEMM and FEM, respectively, which are parallel to the membership function axis.

11.3 Point Kinetic Neutron Diffusion Equation

In this section, the concept of fuzziness along with the stochastic behaviour of the point kinetic neutron diffusion equation has been modelled. The fuzzy stochastic model is investigated numerically by extending the EMM to fuzzy form. The uncertain neutron density and the delayed neutron population are obtained and compared with the Monte Carlo and stochastic principal component analysis (PCA) solutions. Various combinations of fuzzy parameters are considered, and the uncertain neutron density and delayed neutron population are obtained in different cases. Finally, the sensitivity of these fuzzy parameters with stochastic behaviour has been investigated.

11.3.1 Stochastic Point Kinetic Model with Fuzzy Parameters

The point kinetic equation has been modelled in terms of stochastic (Hayes and Allen 2005) by considering the birth and death processes of the neutron and the precursor population.

The coupled deterministic time-dependent equations for the neutron density and the delayed neutron precursors may be represented as

$$\frac{\partial N}{\partial t} = Dv\nabla^2 N - (\Sigma_a - \Sigma_f)vN + \left[(1-\beta)k_\infty \Sigma_a - \Sigma_f\right]vN + \sum_i \lambda_i C_i + S_0 \tag{11.23}$$

$$\frac{\partial C_i}{\partial t} = \beta_i k_\infty \Sigma_a vN - \lambda_i C_i \tag{11.24}$$

for $i = 1, 2, \ldots, m$.

Here, we have considered uncertain parameters, viz. delayed neutron fraction, source and initial condition as fuzzy (Equations 11.23 and 11.24). The fuzzy delayed neutron fraction $\tilde{\beta}$ in α-cut form may be written as $\tilde{\beta} = \left[\underline{\beta}(\alpha), \overline{\beta}(\alpha)\right]$.

These fuzzy parameters are introduced in Equations 11.23 and 11.24, which give

$$\frac{\partial \tilde{N}}{\partial t} = Dv\nabla^2 \tilde{N} - (\Sigma_a - \Sigma_f)v\tilde{N} + \left[(1-\tilde{\beta})k_\infty \Sigma_a - \Sigma_f\right]v\tilde{N} + \sum_i \lambda_i C_i + S_0 \tag{11.25}$$

$$\frac{\partial \tilde{C}_i}{\partial t} = \tilde{\beta}_i k_\infty \sum_a v\tilde{N} - \lambda_i \tilde{C}_i \tag{11.26}$$

for $i = 1, 2, \ldots, m$, where '~' represents fuzzy numbers.

The neutron captures are considered as deaths whereas the fission process is taken as a pure-birth process, where $v-1$ neutrons are born in each fission along with a precursor contribution. The number of new neutrons born in each fission is $(1-\tilde{\beta})v-1$. Let us consider $\tilde{N} = \tilde{f}(r)\tilde{n}(t)$ and $\tilde{C}_i = \tilde{g}_i(r)\tilde{c}_i(t)$, where it is assumed that $\tilde{N}$ and $\tilde{C}_i$ are separable in time and space. Proceeding further and using the technique given in Chapter 10, we get the following uncertain point kinetic equation:

$$\frac{d\tilde{n}}{dt} = -\left[\frac{-\rho+1-\alpha}{l}\right]\tilde{n} + \left[\frac{1-\alpha-\tilde{\beta}}{l}\right]\tilde{n} + \sum_{i=1}^{m}\lambda_i \tilde{c}_i + \tilde{q} \tag{11.27}$$

$$\frac{d\tilde{c}_i}{dt} = \frac{\tilde{\beta}_i}{l}\tilde{n} - \lambda_i \tilde{c}_i, \quad i = 1, 2, \cdots, m \tag{11.28}$$

where

$\tilde{n}(t)$ is the total number of uncertain neutrons

$\tilde{c}_i(t)$ is the total number of uncertain precursors of the ith type at time t

the reactivity $\rho = \dfrac{\Sigma_a(1-k_\infty)+D\nabla^2}{\Sigma_a+D\nabla^2}$

the number of neutrons per fission process is $\alpha = \dfrac{\Sigma_f}{\Sigma_a k_\infty} \approx \dfrac{1}{v}$, $l = \dfrac{1}{v\Sigma_a k_\infty}$ and

$\tilde{q} = \tilde{q}(t) = \dfrac{S_0(r, t)}{\tilde{f}(r)}$

Finally, we obtain the following Ito SDE for one precursor in terms of fuzzy:

$$\frac{d}{dt}\begin{bmatrix} \tilde{n} \\ \tilde{c}_1 \end{bmatrix} = \tilde{A}\begin{bmatrix} \tilde{n} \\ \tilde{c}_1 \end{bmatrix} + \begin{bmatrix} \tilde{q} \\ 0 \end{bmatrix} + \tilde{B}^{(1/2)}\frac{d\vec{W}}{dt} \tag{11.29}$$

where

$$\tilde{A} = \begin{bmatrix} \dfrac{\rho - \tilde{\beta}}{l} & \lambda_1 \\[3mm] \dfrac{\tilde{\beta}_1}{l} & -\lambda_1 \end{bmatrix}$$

$$\tilde{B} = \begin{bmatrix} \gamma n + \lambda_1 c_1 + \tilde{q} & \dfrac{\tilde{\beta}_1}{l}\left(\left(1 - \tilde{\beta}\right)v - 1\right)n - \lambda_1 c_1 \\[4mm] \dfrac{\tilde{\beta}_1}{l}\left(\left(1 - \tilde{\beta}\right)v - 1\right)n - \lambda_1 c_1 & \dfrac{\tilde{\beta}_1^2 v}{l}n + \lambda_1 c_1 \end{bmatrix}$$

$$\gamma = \frac{-1 - \rho + 2\tilde{\beta} + \left(1 - \tilde{\beta}\right)^2 v}{l}$$

$$\vec{W} = \vec{W}(t) = \begin{bmatrix} W_1(t) \\ W_2(t) \end{bmatrix}$$

Here, $W_1(t)$ and $W_2(t)$ are Wiener processes.

Generalizing Equation 11.23, the stochastic point kinetic equation for m precursors becomes

$$\frac{d}{dt}\begin{bmatrix} \tilde{n} \\ \tilde{c}_1 \\ \tilde{c}_2 \\ \vdots \\ \tilde{c}_m \end{bmatrix} = \tilde{A}\begin{bmatrix} \tilde{n} \\ \tilde{c}_1 \\ \tilde{c}_2 \\ \vdots \\ \tilde{c}_m \end{bmatrix} + \begin{bmatrix} \tilde{q} \\ 0 \\ 0 \\ \vdots \\ 0 \end{bmatrix} + \tilde{B}^{(1/2)}\frac{d\vec{W}}{dt} \tag{11.30}$$

where

$$\tilde{A} = \begin{bmatrix} \dfrac{\rho - \tilde{\beta}}{l} & \lambda_1 & \lambda_2 & \cdots & \lambda_m \\[3mm] \dfrac{\tilde{\beta}_1}{l} & -\lambda_1 & 0 & \cdots & 0 \\[3mm] \dfrac{\tilde{\beta}_2}{l} & 0 & -\lambda_2 & \ddots & \vdots \\[3mm] \vdots & \vdots & \ddots & \ddots & 0 \\[3mm] \dfrac{\tilde{\beta}_m}{l} & 0 & \cdots & 0 & -\lambda_m \end{bmatrix}$$

and

$$
\tilde{\tilde{B}} = \begin{bmatrix}
\tilde{\zeta} & \tilde{a}_1 & \tilde{a}_2 & \cdots & \tilde{a}_m \\
\tilde{a}_1 & \tilde{r}_1 & \tilde{b}_{2,3} & \cdots & \tilde{b}_{2,m} \\
\tilde{a}_2 & 0 & \tilde{r}_2 & \ddots & \vdots \\
\vdots & \vdots & \ddots & \ddots & \tilde{b}_{m-1,m} \\
\tilde{a}_m & \tilde{b}_{m,2} & \cdots & \tilde{b}_{m,m-1} & \tilde{r}_m
\end{bmatrix}
$$

Here,

$$
\tilde{\zeta} = \gamma n + \sum_{i=1}^{m} \lambda_i c_i + \tilde{q}
$$

$$
\tilde{a}_i = \frac{\tilde{\beta}_i}{l}\left(\left(1-\tilde{\beta}\right)v - 1\right)n - \lambda_i c_i
$$

$$
\tilde{b}_{i,j} = \frac{\tilde{\beta}_{i-1}\tilde{\beta}_{j-1}v}{l} n
$$

$$
\tilde{r}_i = \frac{\tilde{\beta}_i^2 v}{l} n + \lambda_i c_i
$$

In compact form, Equation 11.30 may be written as

$$
\frac{d}{dt}\tilde{x} = \tilde{\tilde{A}}\tilde{x} + \begin{bmatrix} \tilde{q} & 0 & \cdots & 0 \end{bmatrix}^T + \tilde{\tilde{B}}^{(1/2)}\frac{d\vec{W}}{dt} \tag{11.31}
$$

where $\tilde{x} = \begin{bmatrix} \tilde{n} & \tilde{c}_1 & \tilde{c}_2 & \cdots & \tilde{c}_m \end{bmatrix}^T$.

11.3.2 Case Study

Some authors have investigated the stochastic point kinetic neutron diffusion equation by using different methods, viz. Monte Carlo and Stochastic PCA. As mentioned previously, here we have considered the fuzziness in the parameters of the stochastic differential equation. Following are the crisp parameters used in the problem, $\lambda_1 = 0.1$, $v = 2.5$, $l = (2/3)$ and $\rho = -(1/3)$ for $t \geq 0$. Uncertain parameters in terms of TFN are taken as $\tilde{\beta} = \tilde{\beta}_1 = [0.02, 0.05, 0.08]$, $\tilde{q} = [195, 200, 205]$ and $\tilde{x}(0) = \begin{bmatrix} [390, 400, 410] & [290, 300, 310] \end{bmatrix}^T$.

The hybridization of stochastic and fuzziness has been investigated in various cases depending upon the combination of uncertain parameters, viz. one, two and three (all) parameters as fuzzy. Uncertain point kinetic SDE has been studied by using the developed FEMM (Nayak and Chakraverty 2016). The distribution of randomness in the neutron and the neutron precursor population is depicted in Figures 11.26 through 11.53. The uncertain neutron population has been discussed in Cases 1 through 3 as follows.

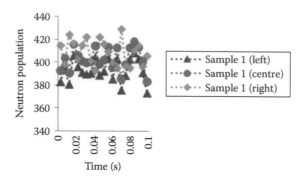

FIGURE 11.26
Initial condition fuzzy (sample 1).

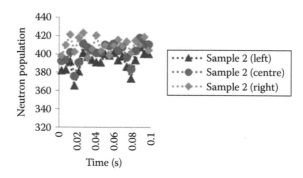

FIGURE 11.27
Initial condition fuzzy (sample 2).

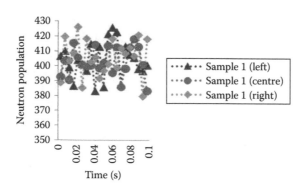

FIGURE 11.28
Source as fuzzy (sample 1).

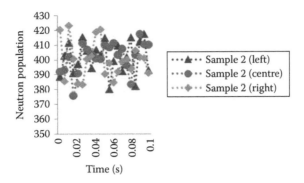

FIGURE 11.29
Source as fuzzy (sample 2).

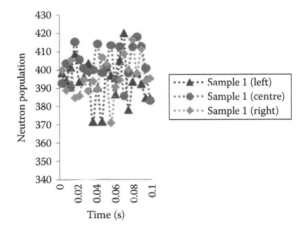

FIGURE 11.30
Neutron precursor constant as fuzzy (sample 1).

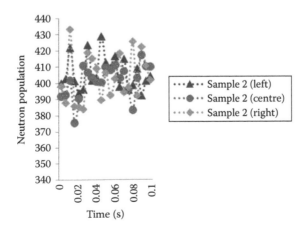

FIGURE 11.31
Neutron precursor constant fuzzy (sample 2).

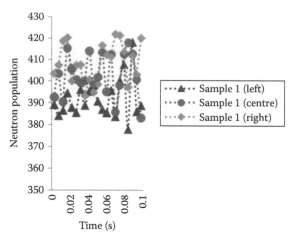

FIGURE 11.32
Initial condition and neutron precursor constant as fuzzy (sample 1).

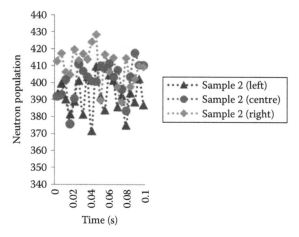

FIGURE 11.33
Initial condition and neutron precursor constant as fuzzy (sample 2).

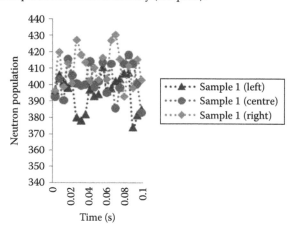

FIGURE 11.34
Initial condition and source as fuzzy (sample 1).

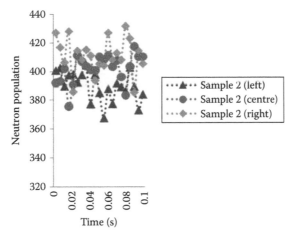

FIGURE 11.35
Initial condition and source as fuzzy (sample 2).

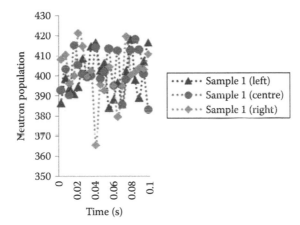

FIGURE 11.36
Neutron precursor constant and source as fuzzy (sample 1).

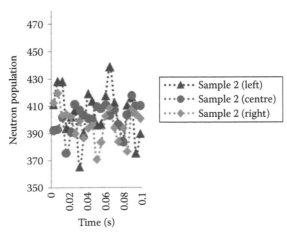

FIGURE 11.37
Neutron precursor constant and source as fuzzy (sample 2).

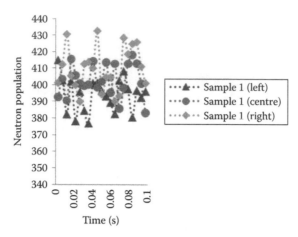

FIGURE 11.38
Initial condition, source and neutron precursor constant as fuzzy (sample 1).

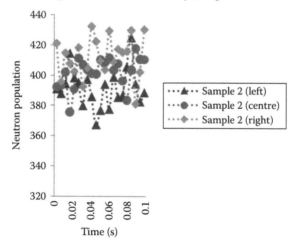

FIGURE 11.39
Initial condition, source and neutron precursor constant as fuzzy (sample 2).

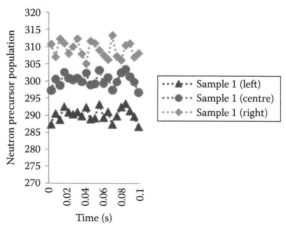

FIGURE 11.40
Initial condition as fuzzy (sample 1).

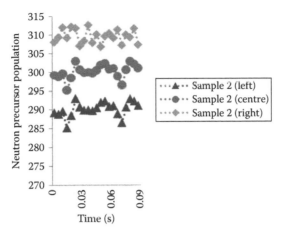

FIGURE 11.41
Initial condition as fuzzy (sample 2).

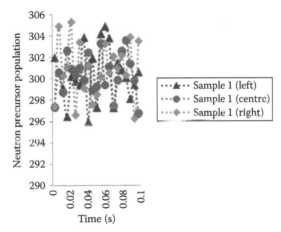

FIGURE 11.42
Source as fuzzy (sample 1).

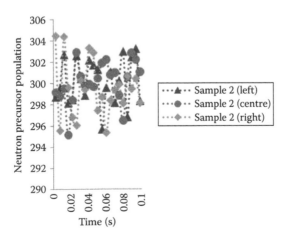

FIGURE 11.43
Source as fuzzy (sample 2).

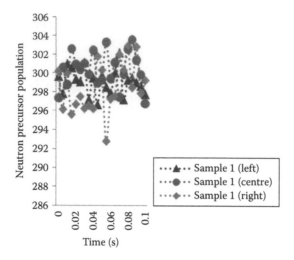

FIGURE 11.44
Neutron precursor constant as fuzzy (sample 1).

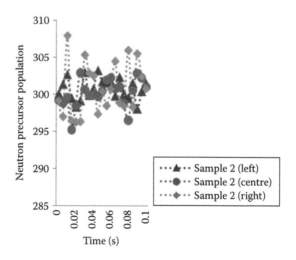

FIGURE 11.45
Neutron precursor constant as fuzzy (sample 2).

Case 1

In this case, only one parameter has been considered as fuzzy and the problem is solved by using the hybrid FEMM (Nayak and Chakraverty 2016) for two samples. In Figures 11.26 and 11.27, the uncertain results of the neutron population are presented, where the initial condition is taken as fuzzy and it is seen that the distribution in the left, right and centre is different. Furthermore, the neutron source is assumed to be fuzzy and the obtained uncertain neutron population has been shown in Figures 11.28 and 11.29. Finally, the neutron

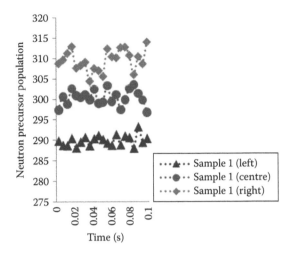

FIGURE 11.46
Initial condition and neutron precursor constant as fuzzy (sample 1).

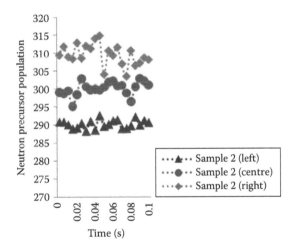

FIGURE 11.47
Initial condition and neutron precursor constant as fuzzy (sample 2).

precursor constant (β) is assumed as fuzzy and the uncertain neutron population is presented graphically in Figures 11.30 and 11.31.

These graphical solutions are given in a tabular form for better visualization of the uncertainty in Table 11.6. From Table 11.6, one may observe that when only the initial condition is taken as fuzzy, the uncertain width is very small for both the samples in various sub-cases, which are shown in Figures 11.26 and 11.27. Hence, we may conclude that when only one parameter is considered as fuzzy, the uncertain initial condition is most sensitive, whereas the uncertain source is least sensitive.

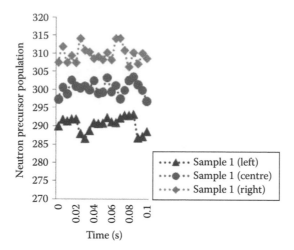

FIGURE 11.48
Initial condition and source as fuzzy (sample 1).

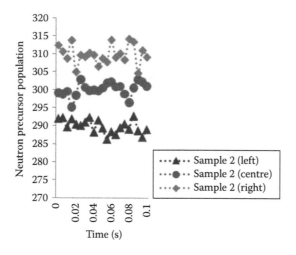

FIGURE 11.49
Initial condition and source as fuzzy (sample 2).

Case 2

Here, two parameters are assumed as fuzzy and various combinations of the two parameters are (1) initial condition and neutron precursor constant, (2) initial condition and source and (3) the neutron precursor constant and source. In Figures 11.32 and 11.33, initial condition and neutron precursor constant are taken as fuzzy and the distribution of the uncertain neutron population has been presented, whereas in Figures 11.34 and 11.35, the initial

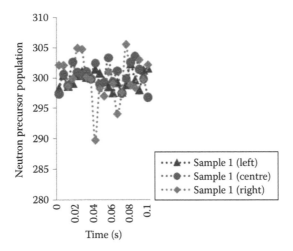

FIGURE 11.50
Neutron precursor constant and source as fuzzy (sample 1).

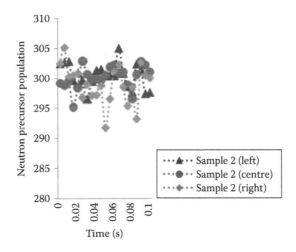

FIGURE 11.51
Neutron precursor constant and source as fuzzy (sample 2).

condition and the source are fuzzy, and the uncertain neutron populations are shown. Finally, the neutron precursor constant and the source are considered as fuzzy and the uncertain results are depicted in Figures 11.36 and 11.37.

The uncertain widths of triangular fuzzy neutron population may be clearly investigated in Table 11.7. We have seen that the combination of the neutron precursor constant and the source as fuzzy has less width than other sub-cases for both the samples. So we may conclude that the combination of the neutron precursor constant and the source is less sensitive than other combinations.

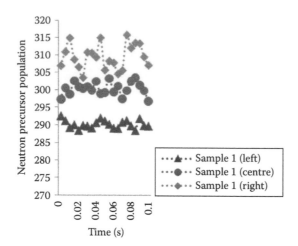

FIGURE 11.52
Initial condition, source and neutron precursor constant as fuzzy (sample 1).

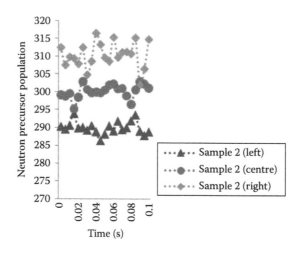

FIGURE 11.53
Initial condition, source and neutron precursor constant as fuzzy (sample 2).

Case 3

Finally, in Case 3, all the parameters, viz. the initial condition, the source and the precursor constant, are taken as fuzzy. The obtained uncertain neutron population for Samples 1 and 2 is shown in Figures 11.38 and 11.39, respectively. Furthermore, the uncertain width of the samples has been reported in Table 11.8.

Similarly, we have investigated the uncertain neutron precursor population of the system and these are discussed in various cases, that is Cases 4 through 6.

TABLE 11.6

Comparison of Neutron Population When Only One Parameter Is Fuzzy

Fuzzy Parameters	Samples	Expectations (Hayes and Allen 2005)	Monte Carlo (Crisp) (Hayes and Allen 2005)	FEMM		Uncertain Width	Stochastic PCA (Crisp) (Hayes and Allen 2005)
				TFN	$\alpha = 0$		
Initial condition	Sample 1	$E(n(2))$	400.03	[392.60, 402.53, 408.18]	[392.60, 408.18]	15.58	395.32
	Sample 2	$E(n(2))$		[391.88, 401.82, 409.65]	[391.88, 409.65]	17.77	
Source	Sample 1	$E(n(2))$	400.03	[402.53, 403.65, 405.68]	[402.53, 405.68]	3.15	395.32
	Sample 2	$E(n(2))$		[399.93, 401.64, 401.82]	[399.93, 401.82]	1.89	
Neutron precursor constant	Sample 1	$E(n(2))$	400.03	[393.87, 396.13, 402.53]	[393.87, 402.53]	8.66	395.32
	Sample 2	$E(n(2))$		[401.82, 403.36, 406.04]	[401.82, 406.04]	4.22	

TABLE 11.7

Comparison of Neutron Population When Only Two Parameters Are Fuzzy

Fuzzy Parameters	Samples	Expectations (Hayes and Allen 2005)	Monte Carlo (Crisp) (Hayes and Allen 2005)	FEMM		Uncertain Width	Stochastic PCA (Crisp) (Hayes and Allen 2005)
				TFN	$\alpha = 0$		
Initial condition and neutron precursor constant	Sample 1	$E(n(2))$	400.03	[391.81, 402.53, 408.70]	[391.81, 408.70]	16.89	395.32
	Sample 2	$E(n(2))$		[391.76, 401.82, 409.50]	[391.76, 409.50]	17.74	
Initial condition and source	Sample 1	$E(n(2))$	400.03	[394.98, 402.53, 410.10]	[394.98, 410.10]	15.12	395.32
	Sample 2	$E(n(2))$		[390.03, 401.82, 411.05]	[390.03, 411.05]	21.02	
Source and neutron precursor constant	Sample 1	$E(n(2))$	400.03	[399.91, 400.92, 402.53]	[399.91, 402.53]	2.62	395.32
	Sample 2	$E(n(2))$		[396.90, 401.82, 404.60]	[396.90, 404.60]	7.7	

TABLE 11.8

Comparison of Neutron Population When All Parameters Are Fuzzy

Fuzzy Parameters	Samples	Expectations (Hayes and Allen 2005)	Monte Carlo (Crisp) (Hayes and Allen 2005)	FEMM		Uncertain Width	Stochastic PCA (Hayes and Allen 2005)
				TFN	$\alpha = 0$		
All fuzzy	Sample 1	$E(n(2))$	400.03	[393.93, 402.53, 409.59]	[393.93, 409.59]	15.66	395.32
	Sample 2	$E(n(2))$		[391.49, 401.82, 411.87]	[391.49, 411.87]	20.38	

Case 4

In this case, only one parameter has been considered as fuzzy and the others are crisp. In Figures 11.40 and 11.41, the initial condition is fuzzy. Here, two samples have been investigated and we see that there is no overlapping between the left, center and right distributions of the uncertain neutron precursor constant. The source term has been taken as fuzzy in Figures 11.42 and 11.43, whereas in Figures 11.44 and 11.45, the neutron precursor constant is considered as fuzzy.

One may see that when the initial condition is taken as fuzzy, we get a larger width for both the samples, whereas for the other problems, viz. when only the source and only the neutron precursor constant are fuzzy, then the uncertain width is less for both the samples and these are presented in Table 11.9. So, we may conclude that the initial condition is more sensitive.

Case 5

In this case, two parameters are taken as fuzzy and the other is crisp. In Figures 11.46 and 11.47, the initial condition and the neutron precursor constant are fuzzy. Again, two samples have been observed and we see that there is no overlapping between the left, center and right distributions of the uncertain neutron precursor constant. In Figures 11.48 and 11.49, the initial condition and the source are considered as fuzzy, whereas the source and the neutron precursor constant are assumed as fuzzy in Figures 11.50 and 11.51.

The uncertain width of the problems, viz. (1) the initial condition and the neutron precursor constant are fuzzy, (2) the initial condition and the source are fuzzy and (3) the source and the neutron precursor constant are fuzzy, for both Samples 1 and 2 are incorporated in Table 11.10. We may observe that the uncertain width for (3) is the least. So we may conclude that the fuzziness in the combination of the source and the neutron precursor constant is less sensitive in comparison with the other two combinations.

Case 6

Here, all three parameters are considered as fuzzy, and the left, right and centre distributions of the uncertain neutron precursor population have been shown in Figures 11.52 and 11.53. Furthermore, in Table 11.11, uncertain widths of the neutron precursor population for the samples are presented.

TABLE 11.9

Comparison of Neutron Precursor Population When Only One Parameter Is Fuzzy

Fuzzy Parameters	Samples	Expectations (Hayes and Allen 2005)	Monte Carlo (Crisp) (Hayes and Allen 2005)	FEMM TFN	FEMM α = 0	Uncertain Width	Stochastic PCA (Crisp) (Hayes and Allen 2005)
Initial condition	Sample 1	$E(c(2))$	300.00	[290.33, 300.32, 309.17]	[290.33, 309.17]	18.84	300.67
	Sample 2	$E(c(2))$		[290.11, 300.10, 309.62]	[290.11, 309.62]	19.51	
Source	Sample 1	$E(c(2))$	300.00	[300.32, 300.32, 300.85]	[300.32, 300.85]	0.53	300.67
	Sample 2	$E(c(2))$		[299.48, 300.10, 300.16]	[299.48, 300.16]	0.68	
Neutron precursor constant	Sample 1	$E(c(2))$	300.00	[298.44, 299.10, 300.32]	[298.44, 300.32]	1.88	300.67
	Sample 2	$E(c(2))$		[300.10, 300.69, 300.73]	[300.10, 300.73]	0.63	

TABLE 11.10

Comparison of Neutron Precursor Population When Two Parameters Are Fuzzy

Fuzzy Parameters	Samples	Expectations (Hayes and Allen 2005)	Monte Carlo (Crisp) (Hayes and Allen 2005)	FEMM TFN	FEMM α = 0	Uncertain Width	Stochastic PCA (Crisp) (Hayes and Allen 2005)
Initial condition and neutron precursor constant	Sample 1	$E(c(2))$	300.00	[289.86, 300.32, 309.50]	[289.86, 309.50]	19.64	300.67
	Sample 2	$E(c(2))$		[290.08, 300.10, 309.43]	[290.08, 309.43]	19.35	
Initial condition and source	Sample 1	$E(c(2))$	300.00	[290.43, 300.32, 309.67]	[290.43, 309.67]	19.24	300.67
	Sample 2	$E(c(2))$		[289.83, 300.10, 309.73]	[289.83, 309.73]	19.9	
Source and neutron precursor constant	Sample 1	$E(c(2))$	300.00	[299.68, 299.76, 300.32]	[299.68, 300.32]	0.49	300.67
	Sample 2	$E(c(2))$		[298.91, 300.10, 300.38]	[298.91, 300.38]	1.47	

TABLE 11.11

Comparison of Neutron Precursor Population When All Parameters Are Fuzzy

Fuzzy Parameters	Samples	Expectations (Hayes and Allen 2005)	Monte Carlo (Crisp) (Hayes and Allen 2005)	FEMM TFN	FEMM α = 0	Uncertain Width	Stochastic PCA (Crisp) (Hayes and Allen 2005)
All fuzzy	Sample 1	$E(c(2))$	300.00	[290.16, 300.32, 309.51]	[290.16, 309.51]	19.35	300.67
	Sample 2	$E(c(2))$		[289.90, 300.10, 310.22]	[289.90, 310.22]	20.32	

Bibliography

Bart, L. S. and Hoogenboom, J. E. 2013. Dynamic Monte Carlo method for nuclear reactor kinetics calculations. *Nuclear Science and Engineering* 175:94–107.

Black, F. and Scholes, M. 1973. The pricing of options and corporate liabilities. *Journal of Political Economy* 81:637–654.

Caro-Corrales, J., Cronin, K., Abodayeh, K., Gutierrez-Lopez, G. and Ordorica-Falomir, C. 2002. Analysis of random variability in biscuit cooling. *Journal of Food Engineering* 54:147–156.

Chakraverty, S. and Nayak, S. 2013. Non probabilistic solution of uncertain neutron diffusion equation for imprecisely defined homogeneous bare reactor. *Annals of Nuclear Energy* 62:251–259.

Glasstone, S. and Sesonke, A. 2004. *Nuclear Reactor Engineering*. CBS Publishers and Distributors Private Limited, New Delhi, India.

Hayes, J. G. and Allen, E. J. 2005. Stochastic point kinetic equations in nuclear reactor dynamics. *Annals of Nuclear Energy* 32:572–587.

Hetrick, D. L. 1971. *Dynamics of Nuclear Reactors*. University of Chicago Press, USA.

Higham, D. J. 2001. An algorithmic introduction to numerical simulation of stochastic differential equations. *SIAM Review* 34:525–546.

Higham, D. J. and Kloeden, P. 2005. Numerical methods for nonlinear stochastic differential equations with jumps. *Numerische Mathematik* 101:101–119.

Kim, J. H. 2005. On fuzzy stochastic differential equations. *Journal of Korean Mathematical Society* 42:153–169.

Kloeden, P. and Platen, E. 1992. *Numerical Solution of Stochastic Differential Equations*. Springer, Berlin, Germany.

Malinowski, M. T. and Michta, M. 2011. Stochastic fuzzy differential equations with an application. *Kybernetika* 47(1):123–143.

Nayak, S. and Chakraverty, S. 2013. Non-probabilistic approach to investigate uncertain conjugate heat transfer in an imprecisely defined plate. *International Journal of Heat and Mass Transfer* 67:445–454.

Nayak, S. and Chakraverty, S. 2016. Numerical solution of stochastic point kinetic neutron diffusion equation with fuzzy parameters. *Nuclear Technology* 193(3):444–456.

Ogura, Y. 2008. On stochastic differential equations with fuzzy set coefficients. In: Dubois, D., et al. (eds.), *Soft Methods for Handling Variability and Imprecision*, pp. 263–270. Springer, Berlin, Germany.

Oksendal, B. 2003. *Stochastic Differential Equations: An Introduction with Applications*. Springer-Verlag, Heidelberg, Germany.

Platen, E. 1999. *An Introduction to Numerical Methods for Stochastic Differential Equations. Acta Numerica* 8:197–246.

Rumelin, W. 1982. Numerical treatment of stochastic differential equations. *SIAM Journal on Numerical Analysis* 19:604–613.

Sauer, T. 2012. *Numerical Solution of Stochastic Differential Equations in Finance*. Springer, USA.

Zadeh, L. A. 1965. Fuzzy sets. *Information and Control* 8:338–353.

Zimmermann, H. J. 1991. *Fuzzy Sets Theory and Its Applications*. Kluwer Academic Press, Dordrecht, the Netherlands.

Index